# AVIONICS CERTIFICATION

## A Complete Guide to DO-178 (Software) DO-254 (Hardware)

Vance Hilderman
and
Tony Baghi

www.avionics.com

Avionics Communications Inc.
Leesburg, VA

ISBN 978-1-885544-25-4


First edition 2007

Library of Congress Cataloging-in-Publication Data

Hilderman, Vance.
Avionics certification : a complete guide to DO-178 (software), DO-254 (hardware) / Vance Hilderman, and Tony Baghi.
p. cm.
ISBN 978-1-885544-25-4
1. Avionics--Computer programs--Certification--United States. 2. Guidance systems (Flight)--Certification--United States. 3. Avionics--Computer programs--Standards--United States. 4. Guidance systems (Flight)--Standards--United States. I. Baghi, Tony. II. Title.

TL695.H55 2007
629.135--dc22
2007000314

---

The information in this book is provided solely as guidance to the certification of aviation safety systems. To determine compliance for a specific system or product, the reader should consult a qualified Designated Engineering Representative and confer with the appropriate regulatory agency.

Printed in the United States of America

Publisher:
Avionics Communications Inc.
P.O. Box 2628
Leesburg, VA 20177 USA
Tel: 1-703/777-9535
www.avionics.com
e-mail: len@avionics.com

# About the Authors

**Vance Hilderman** spent over 25 years designing and testing aerospace systems and software. He began his career at Hughes Aircraft after graduating with BSEE and MBA degrees from Gonzaga University and an MS Computer Engineering degree from the University of Southern California as a Hughes Fellow. His experience in aerospace includes over 40 different aircraft and 150 systems.

With funding from Rockwell Collins Avionics in 1990, Vance founded a company that became the world's largest avionics software services organization. While completing several hundred programs for all fifty of the world's largest and best-known avionics companies, he introduced the first public training for DO-178. Along with his colleagues, he has trained nearly 6.000 industry professionals in DO-178 and DO-254; more than all other private and government-sponsored DO-178 seminars combined.

Today, Vance resides in San Diego with his three children and is an advisor to HighRely, a company he co-founded to meet the growing need for avionics and safety-critical software services and consulting. Vance is also the founding director of Weekend Warrior Men, a non-profit entity. He can be contacted at: hilderman@sprynet.com.

**Tony Baghai** has worked in the aerospace industry for close to 20 years on the design, development and certification of systems and software. With a BS degree in Mechanical Engineering and an MS in Aeronautical Engineering, he had a long and distinguished career at Boeing Commercial Aircraft. During that time, Tony became one of the youngest Designated Engineering Representatives (DER) ever to earn FAA certification for Systems and Level A Software. During fourteen years as a DER he has traveled the world, becoming well-known as an avionics certification expert.

Tony managed the certification division of one of the largest avionics services company in the world, where he worked with nearly every major avionics manufacturer. He is co-founder of HighRely, along with Reza Madjidi and co-author Vance Hilderman. Educated in England and the United States, he makes his home in Scottsdale, Arizona, with his family. He may be reached at info@highrely.com.

**HighRely** has grown rapidly in avionics software services and training in recent years. Software developed or certified by its engineers is aboard civil and military aircraft in such systems as flight control, electrical power, landing gear, navigation, communications, terrain avoidance and electronic instruments (EFIS), as well as ground-based and space systems. Specific programs include:

*Boeing C-17 - upgrade using DO-178B –Inertial Reference Unit*
*Boeing 767 - Global Tanker Aircraft Actuator Control Unit (ACU)*
*Boeing 777 - Airplane Information Management System (AIMS), Maintenance Access Terminal (MAT)*
*Boeing 777 - Thrust Reversers and Power Supply Management*
*Joint Strike Fighter - Fuel Flow*
*Boeing 787 – Multiple programs*
*Bombardier-Global Express - Secondary Power Distribution System*
*Rockwell Collins - ELS (Electronic Library System)*
*Boeing 747, 757, 767 – Multiple TSO's Displays and Navigation, EFIS (Electronic Flight Information System)*
*Boeing 737 - Smoke Detection System*
*General Aviation - Terrain Avoidance and Navigation Systems*
*Bell Helicopter – various rotorcraft systems*
*General Atomics - Predator UAV systems (Unmanned Aerial Vehicle)*
*Israeli and South African military airborne systems*
*Japanese Air Defense Systems*
*Saab Aerospace – multiple systems on Grippen*
*General aviation – multiple systems for multiple manufacturers*
*P3 Orion - Antisubmarine Warfare Systems*
*Airbus A380 - Multiple programs*
*Airbus 400M – Multiple programs*

A description of the company's engineering activity may be seen at www.highrely.com. Private training in DO-178B, DO-178, DO-254, avionics project management, systems engineering and embedded software development are available from HighRely via info@highrely.com. The company also produces a set of packages consisting of products and services for avionics software and hardware developers called JumpCert-178 and JumpCert-254, respectively.

The authors will donate all royalties from the sale of this book to non-profit organizations affiliated with aircraft safety.

# Contents

# 1. Introduction

DO-178B was developed by the commercial avionics industry and RTCA (a Federal Advisory Committee) to establish guidelines for avionics software developers and government certifying agencies. The first version, DO-178, covered the basic avionics software life cycle, followed by DO-178A, which added criticality levels and emphasized software component testing to obtain quality. The current version, a complete re-write, created DO-178B, brought improvements in software quality by adding more planning, created data items rather than documents, allows a variety of software development practices, permits commercial off the shelf (COTS) products and tools, continuous quality monitoring and verification and testing in real-world conditions.

DO-178B is a guideline; but in reality a strict requirement. At merely 100 pages in length, it must be all things to all people, which makes it quite broad in nature. Nevertheless, it requires in-depth understanding of intent. To be properly used, a DO-178-B certification must be accompanied by much documentation and end-system case studies.

DO-254 (also known as Eurocae ED-80) deals with hardware. It is a formal avionics standard for design assurance of airborne electronic hardware. DO-254 provides information from project conception, planning, design, implementation, testing and validation, including tool qualification.

The two documents, DO-254 and DO-178B, are quite similar, both having been written by people with formal software process expertise. Until recently, avionics hardware certification did not require the same strict standards as did software. But avionics systems are comprised of both hardware and software, each having near-equal effect on airworthiness. Today, most avionics projects come under a DO-254 certification or a compliance mandate along with the typical DO-178B mandate.

DO-178B offers many benefits beyond certification, such as verifiable software quality, higher reliability, consistency, greater re-usability, lower life cycle costs, decreased maintenance, faster hardware integration and greater probability of catching failures within the development process.

Over the years we've seen people take unnecessary risks along the road to certification. They may, in fact, do *too* much work and neglect key artifacts or steps which cause cost overruns averaging 40%---and still not meet the true intent of DO-178B. Each of these pitfalls is discussed in detail throughout this book---as well as how to avoid them. But first, let's look at several frequently-asked questions heard from people about to start a DO-178 certification program.

### Can you apply DO-178B reverse engineering to existing software?

Yes. While DO-178B applies to new, custom software, there are provisions to apply reverse-engineering to previously developed software, and preserve most of the completed work.

.

### What is DO-178B Tool Qualification?

Software development requires many tools for design, code generation, compilers/linkers, libraries, test and structural coverage. In DO-178B, tool qualification pertains to development and verification/testing tools. Different qualification criteria apply to each, and most tools do *not* need to be qualified. When it is required, tool qualification uses a subset of DO-178B and ensures the proper operation of the tool.

### What is DO-178B Gap Analysis?

This evaluates your current software engineering process and artifacts, as contrasted to those required by DO-178B. While DO-178B covers custom-developed avionics software, it recognizes that previously developed software can be DO-178B compliant or certified. In many cases, particularly for military avionics software, DO-178B *compliance* is used instead of *certification.* DO-178B compliance is near-certification, but may not require FAA involvement.

Gap Analysis is typically performed by trained consultants or Designated Engineering Representatives (DERs). The result is a Gap Analysis that assesses all software processes and artifacts. It provides details for filling the gap to meet DO-178B certification or compliance requirements.

### What is MCDC?

Modified (or Multiple) Condition Decision Coverage means that every point of entry and exit in the program is invoked at least once, every condition in a decision takes all possible outcomes at least once, and each condition is shown to affect that outcome independently.

The key to successful MCDC analysis and testing is to analyze each source code construct for potential MCDC applicability, then developing sufficient test cases to ensure that each condition in that construct is independently verified per the MCDC definition. Today, most MCDC analysis/testing is done with the assistance of DO-178B-qualified structural coverage tools.

### What is dead code?

This is executable (binary) software that will never be executed during runtime operations. D0-178B generally does not allow dead code, which must be removed. Dead code does not trace to any software requirements, hence does not perform any required functionality. Unreferenced variables or functions which are not called elsewhere in the program are usually removed via the compiler or linker. Since they are not present in the binary executable load image, they are not dead code per DO-178B.

### What is deactivated code?

This is executable (binary) software that will not be executed during runtime operations of a software configuration or version within a particular avionics box. However, the code may be executed during maintenance or special operations, or executed within a different configuration or future version. Unlike dead code, deactivated code may be left in the source and binary baseline.

### What is DO-178B requirements traceability?

This is correlating individual requirements to the design, code and verification/test elements affiliated with implementing and verifying each requirement. Requirements traceability can be many-to-one and one-to-many.

### Which language is best for avionics software?

High order languages (requiring a compiler with complex syntax construction capabilities) are strongly preferred because they are safer. Safe avionics software? DO-178B emphasizes code consistency, visibility, determinism, defensive coding, robustness, design traceability, software peer reviews per detailed checklists, thorough testing via structural coverage and real-world asynchronous testing.

As a result, avionics code is best written in Ada, C and C++. With all languages, a safe subset should be used. Ada was the *defacto* avionics language standard, and Ada95 improved object oriented capabilities. However, the tide is behind C and C++; not because of superiority, but the wider availability of development tools, and engineers able to develop real-time embedded C and C++.

### Which Configuration Management (CM) tools are best?

DO-178B requires configuration management of all software lifecycle artifacts, including requirements, design, code, tests and documentation. However, DO-178B does not require specific tools, not even for avionics configuration management. Thus, configuration management can be performed manually, even with a purely paper-based system. Virtually all avionics and DO-178B software projects are better served, however, with a configuration management tool. Simple tools (free or low-cost: $0-$200/user) provide basic software version control, check-in/check-out and document management. Higher cost tools provide more complexity and automation. No commercially available CM tool known to us, however, performs all required DO-178B configuration management process steps. In particular, data security, offsite backups, peer reviewing each change and ensuring no unwarranted changes, are all DO-178B configuration management steps typically performed outside the scope of an avionics configuration management tool, but can be achieved via augmentation of the process.

**What is a DO-178B checklist?**
Checklists track DO-178B compliance. Overview checklists are provided in the back of DO-178B; however, these are not acceptable checklists since actual checklists must be developed (or procured) for each project on a case-by-case basis. Numerous third parties provide DO-178B and DO-254 checklists covering all phases of planning, development, correctness, quality assurance and Designated Engineering Representative (DER) activities.

**What is DO-178B Independence?**
This is a separate development and review authority applied to different DO-178B lifecycle steps. Development refers to origination of a required artifact (requirements, design, code, test, etc). Review authority refers to an individual tasked with the review of correctness and DO-178B compliance review of that artifact.

**What is criticality level?**
There are five criticality levels, with Level A the most critical and Level E the least. This is based on the contribution of software to potential failure conditions. Each avionics function or system has one defined criticality level (which must be approved by the certification agency); however, different components within a system can have differing criticality levels. The higher the criticality level, the greater the amount of software development effort.

**What is DO-178B Level A?**
At this level, anomalous software behavior would result in a catastrophic failure typified by total loss of life. ***Approximately 20-30% of avionics systems, and 40% of software code, must meet Level A.***

**What is Level B?**
This is where anomalous software behavior would contribute to a hazardous/severe-major failure, typified by some loss of life. ***Approximately 20% of avionics systems and 30% of software code must meet DO-178B Level B criteria.***

**What is Level C?**
This is where anomalous behavior would contribute to major failure for the aircraft, typified by serious injuries. ***Approximately 25% of avionics systems and 20% of software code must meet DO-178B Level C criteria.***

### What is Level D?

This is software whose anomalous behavior would contribute to a minor failure for the aircraft, typified by minor injuries. ***Approximately 20% of avionics systems and 10% of avionics software code must meet DO-178B Level D.***

### What is Level E?

This is software whose anomalous behavior would contribute to a failure with no effect on aircraft operational capability or pilot workload. There would be no impact on passenger or aircraft safety. ***Approximately 10% of avionics systems and 5% of software code must meet DO-178B Level E.***

The amount of Level E source code is increasing today because of passenger entertainment and Internet communications. It is likely that their criticality levels will increase due to integration with other, more critical, avionics systems.

### What is DO-178B tool qualification?

This process evaluates software development and verification tools to determine if formal qualification is required. There are two types: development tool qualification and verification tool qualification. DO-178B development tools provide outputs which are actually present in the embedded operational avionics software. Such tools must apply DO-178B software life cycle aspects to ensure integrity. Verification tools assist DO-178B verification. Tools which meet these criteria, and which reduce, automate or replace process steps cited by DO-178B, must be qualified.

### What is structural coverage?

This pertains to proof that formal software verification test cases fully cover software structures (conditions and paths). DO-178B structural coverage is not required for Level E and Level D software; it is required in increasing degrees for Level C, Level B and Level A.

### What is DO-178B certifiability?

This designates that an avionics component can meet a subset of DO-178B, with the remaining certification requirements to be subsequently achieved. DO-178B certification pertains to individual systems, and thus requires all software components of a system to be completed, with each component and the system fully meeting all DO-178B requirements. However, in the absence of a completed system, an individual software component (RTOS, graphics library, communications protocol, etc.) can be designated certifiable by subjecting that *component* to all DO-178B requirements.

# 2. Real World of DO-178B

It's surprising how many times we, in our role as consultants, receive a call that sounds something like:

"We are working on a DO-178B project and need help."
"Great, how much do you know about DO-178B?"
"Nothing."
"Where are you in the project?"
"We just won the proposal."
"OK, you submitted a proposal and had no idea about DO-178B. Is it a level-of-effort proposal or fixed price?"
"Well, it's fixed price..... "

This now-familiar conversation is straight from "The School of Wishful Thinking"----which has many students, but no graduates. It is also why this book was written

A DO-178B certification often adds 20 to 40 percent to the cost of a project. Yet, many pitfalls along the way can quickly raise that to between 70 and 200 percent. Fortunately, the pitfalls can be avoided by looking at how DO-178 works in the real world. It not only avoids wasting time, money and productivity, but keeps software maintainable, reusable and cost-effective.

To look at DO-178 close up, we'll draw on experience gained from over 300 certifications completed during the last **seven** years, including programs now flying on major commercial airframes and military aircraft.

Many comments have been made about DO-178. Let's take a look at them and dispel some of the myths:

DO-178B contains "considerations," not requirements.
It was developed by committee, so it's all things to all people.
It applies to huge and tiny systems alike.
It applies to the most critical, Level A, systems and lowest Level D.
DO-178B is vague and too expensive.

Consider the first point---about DO-178 containing "considerations," not requirements. Nowhere, in fact, does DO-178B say "requirement." It never says "Thou shalt do this." The reason is, 178B was developed by committee, and committees must create documents that are all things to all people, applying equally to large and small systems. If there are 50 lines of code in a simple

> **Objectives**
>
> Understand DO-178B in the real world
>
> Avoid wasting time & money
>
> Maximize productivity while still complying with DO-178B
>
> Leverage experience gained during 300 avionics certifications
>
> Apply industry best practices, tools and products

driver, for example, to turn on a light in an airplane lavatory, those lines must comply with DO-178B---just the same as for an Aircraft Information System with 1.2 million line of code.

Next there is the complaint that DO-178B is vague. And that is true! In spite of improvements, the document *is* vague, and not very clear or easy to navigate. It runs only one hundred pages (and ten pages are the names of people who worked on it). But this is offset by the fact there has never been a successful lawsuit over interpreting requirements in this document. We all know cases that make the headlines---about the grandmother suing over spilling hot coffee in her lap, dogs that bark all night or the neighbor's tree falling into your yard. But no-one has ever challenged DO-178B. Built into the document is an effective anti-lawyer, virus-proofing system. And the reason is that nothing is *required!* If there are no requirements, though, why waste time reading this book? The reason is simple; you won't get certified without applying DO-178B.

If the FAA (or other certifying agency) made it *required*, here's what could happen. Let's say a company develops an operating system and has it certified as software to automatically land an airplane in zero-visibility weather. When the airplane crashes the company could say, "We did nothing wrong, we followed the *required* standard." So whose fault is it? Lawyers might argue that the standard is wrong---because it didn't cover some critical issue---and that caused the crash. They'd be absolutely right. But since DO-178B is not a set of *requirements*, you cannot sue over it. The document is only a suggestion or guideline---but you need it to get certified.

## DO-254 on Hardware

For over 20 years, DO-178 guided software development in aircraft. As systems advanced, the need arose to examine hardware, as well. As a result, a sister document---DO-254---emerged to guide certification of complex electronic hardware. The two documents are similar in many respects because they were written by the same group. DO-254 is software- and process-centric, dealing with embedded code within such hardware devices such as PLD's, FPGA's, ATCL's and BACL's. (Acronyms and abbreviations throughout this book are spelled out in a glossary at the end.)

When faced with the need to certify hardware, the FAA didn't have the engineering resources because the agency had never done such a certification. To fill this gap, an RTCA software committee was convened to write the hardware document, which resulted in DO-254. The document is mainly cut-and-paste from DO-178B and the two are about 80 to 85% identical. If a system has logic devices, ASICs or FPGAs, for example, they come under DO-254. Any item containing custom logic, via silicon or software, must be certified to the same rigor.

### Hardware Accomplishment Summary

| | |
|---|---|
| **System Overview** | **Hardware Overview** |
| Certification Considerations | Hardware Design Life Cycle Description |
| Hardware Life Cycle Data | Previously Developed Hardware |
| Additional Considerations | Alternative Methods |

This document must also identify differences from the approved PHAC and contain identification information for the hardware, the change history and status of the hardware and, finally, a compliance statement.

## Testing

You will spend more money on testing than on software development, particularly for Levels A, B, and C. Testing includes four types; functional, normal range, robustness and structural coverage. An interesting sidelight is that one company, producer of a widely used operating system, can claim *partial* DO-178B compliance. The software has almost *30 million* lines of code and over half have been tested. But how many lines of that code have been structurally covered and proven correct? It's over a million, which is quite an accomplishment. *But what about the other 29 million?* Clearly, the operating system is not DO-178B compliant. The testing types we cover here are different from anything done in the military world, in the medical industry or in SEI.

## Design Data and Control Flow

We will also discuss design data and control flow. They are the reason we once declined a request to certify a well-known operating system to Level C in a new very light jet (VLJ). The software was intended to drive electronic displays on the flight deck. We had to tell the manufacturer, "Sorry, you're going down the wrong path." We pointed to the system, which had two displays for redundancy but they had identical software. If a software anomaly disables one, what's going to happen in the other? The same problem. They didn't take our advice and we didn't get the follow-on work, but vindication occurred the following year when the FAA made them re-work the entire avionics cockpit architecture. The airplane was delayed two years at a developmental cost increase of over $30M. Still, it's a great new plane and "wishful thinking" could have been easily avoided by listening to professional advice and not taking short-cuts. DO-178B is about "determinism" and common sense; both required in heavy doses.

## Tools

Tool qualification is a key aspect of DO-178B. Remember, DO-178B is like a chain; an avionics system is only as strong as its weakest link. Where does a chain break? At the weakest link. But DO-178B (and DO-254) is one of the few standards in the world that allows five different strengths of steel in each chain; the five criticality levels we'll describe throughout this book. Obviously, you could make every avionics system comply with the strongest standard, Level A, and be done with it. But that would greatly increase the cost of the airplane, maintenance, upgrades and spares. Is this in the best interest of the aviation industry? Not if airlines can't afford airplanes and passengers can't buy tickets. So DO-178B has different levels of criticality, depending on the contribution of that link (system) to overall chain (aircraft) safety.

Regarding tools; as one of our certification experts sums it up; "A fool with a tool is simply a more powerful fool." Major mistakes are often made in tool qualification because people qualify them unnecessarily. A compiler, for example, doesn't need qualification. But what about a structural coverage tool for testing and analyzing paths you hit? That little item does require qualification unless you manually re-verify every branch decision, which would take many person-years. Some tools need to be qualified, while many don't. We'll show how to quickly determine this.

There are two kinds of tools, development and verification. The difficult one is the development tool; the verification tool is easy to qualify, requiring only a few person-weeks if it's been qualified before.

## "Guilty until proven innocent"

In American and European law, you are innocent until proven guilty. In DO-178B certification it's the opposite; you are guilty until proven innocent! By observing the intent of the document, keeping good records and following a careful plan, you will get certified at the end of the project. For this reason, it's impossible to learn DO-178B "by the book"---because the document is too vague. After many years we've learned the best approach is teaching software engineers how to think like the FAA.

## "Considerations", not "requirements"

Let's take a closer look at the formal title of the document:

**RTCA DO-178B**
**Software**
**Considerations**
**in Airborne Systems**
**and**
**Equipment Certification**

Note that the word "Requirements" isn't there. By using "Considerations" instead, the document is loosened to the point where it might even be considered vague. But the document works well because it was driven by goals of companies within the industry. Seventy-five percent of all participants came from aerospace, so the document works better than other standards. It's not perfect, but is practical because industry is concerned about schedule, cost and risk.

Look at other standards and you find that people often don't follow them because they're convoluted and difficult. So the idea was to write reasonable rules----and DO-178B is reasonable because the industry helped write it.

There are many compromises, which is why it's a guideline, not a recipe book. Nowhere in DO-178B does it say how to *do* something but says what *needs* to be done. You also have a method of proposing something different, known as AMC, for Alternative Means of Compliance. As long as you can show, theoretically and practically, that it's as good or better, you can do it.

Before taking that route, you probably want to talk to an expert or DER. A company in Canada, for example, used eight transputers for loading purposes and got certified to Level B by alternative means. So it can be done. Consider +DO-178B as a discussion flow, not a rigid prescription. In practice, it really is the golden rule.

# 3. Planning the Project

Let's look at how the project flows. The intent of DO-178B is driven by SAE standard ARP4761 *(Guidelines and Methods for Conducting the Safety Assessment Process on Civil Airborne Systems and Equipment).* It's a safety assessment to determine the criticality level----A, B, C or D--- and architectural input.

DO-178B may become confusing without ARP4761 because that's where system level requirements and safety assessments (including Design Assurance Levels) come from. Some people say, "It doesn't talk about developing good software requirements. How do I do that?" The answer is in these two documents. Nothing is more important for a system's cost-effectiveness and quality than "good" requirements.

**DO-178B Layout**

Overview
System Aspects
Lifecycle
Planning Process..................*Planning*
Development Process...........*Development*
Verification............................ *Correctness*
Configuration Mgmt............. "
Quality Assurance............... "
Certification Liaison)............ "
Overview of Aircraft And Engine Certification
Data and Considerations
A. Objectives by Certification Level

S16

We see many companies with great planning documents but, surprisingly, they don't necessarily have good projects. That's because their planning documents are too good! They have no rips or folds in any of the pages---just a little bit of dust---because they were put on a shelf and not followed. Remember that with DO-178B we're guilty until proven innocent. At the end of the project we must not only prove we had good planning documents, but that we followed every word. And that's what it's all about.

Occasionally, we are asked, "What's the fewest number of people needed for a DO-178B project?"

We reply, "That's an interesting question, why do you ask?"

"Well, I just started my company and it's just me."

"Ok, here's the answer..."

You need a safety expert. You need somebody to write system requirements. You need software and hardware requirements. You need a configuration management systems person, a QA person and a DER. You need an architect for the software. Forget about hardware, for now. You need a software developer, software tester and, with that many people, you need a manager. So you need 11 people. Right? No, it's not right.

You need a DER, but you can rent him. Even mid-sized companies don't have their own. You need someone for Quality Assurance. What about those nine other titles? Just two more; a doer and a reviewer. So there's the DER plus three. Who's the most important? It's the QA person because he initiates the DO-178B conformance and decides where to find the DER. But the doer and reviewer are all you need.

Can you have two software developers on the project, plus QA and a DER? Yes---but this raises the question of whether software developers should review their own work, or write code to their own requirements. The answer is no. Can they follow their own software development processes without following the test processes? Again, the answer is no.

One aspect of DO-178B is that it looks like a big beast requiring a dozen people to feed it, but it really doesn't. Quality Assurance is very important. QA is in charge of making sure we follow the plan and maps DO-178B word-for-word so everything is covered from criticality level to audits to prove we're innocent along the way.

You can view your DO-178B project via three key processes. In the next figure, there's a lot of meaning in how these items are positioned and given relative size.

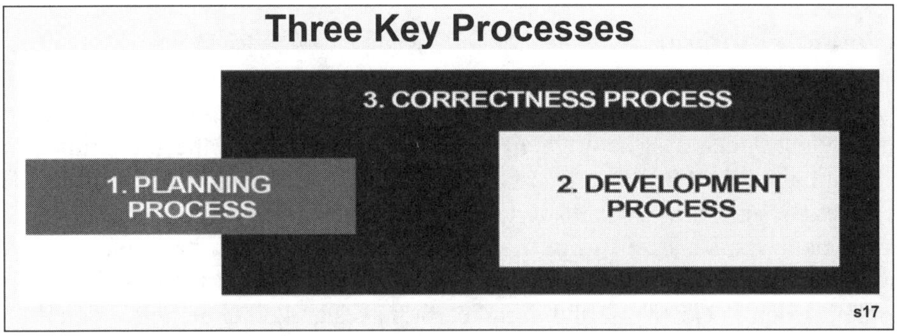


Planning precedes everything, so it's positioned at the left, before anything else. It covers five plans, plus the standards. This is followed by the Development Process. But in the background---much larger than anything is the Correctness Process, which includes testing, quality assurance, configuration management, certification liaison and coordinating with the FAA. It's a major cost driver, but a key process.

**Key Attributes**

Detailed planning

Five criticality Levels (A, B, C, D, E)

Consistency & determinism

Traceability to detailed requirements

Path testing

*"Guilty until proven innocent"*

*s18*

DO-178B is about consistency and determinism. Think of the "D" in DO-178B as standing for "determinism". The priority is exact cause and effect. That's what DO-178B is about.

You also need complete top-to-bottom traceability to show that everything specified is fully implemented. Then you need bottom-to-top traceability for a good reason; to show that only what was specified is included.

## Detailed Planning

There must be plans before starting any development activity. Imagine developing ideas for building a house. One of the first is an architect to lay out the drawings, then a structures engineer to plan and write the blueprints. You can't simply ask, "Build a five-bedroom house with four bathrooms." The architect doesn't know your lifestyle and preferences. He could locate five small bedrooms in the basement---but this is not exactly what you had mind. So you advise the architect on your preferences and he implements them while solving problems such as electrical, plumbing, earthquakes and other technical details.

Next is the professional engineer who has experience with your local approval process. After his review, he stamps the plan acceptable. That's also the role of the DER. Now the building plans can be submitted to the local authority and the reply could be, "Yes. We approve it." Or they might redline certain items. If you start building without preplanning, you are likely to fail---or even be compelled to tear down and rebuild to spec.

For software, you need plans that start with the question; "How will I meet the intent and rules laid out by DO-178B? " You can sum it up this way:

### Detailed Planning

Plans must precede development

Plans must address every aspect of DO-178B

Must provide proof that plans are followed

Plans address what, when, and who ....and a small amount of "how"

Plans are typically written, accepted and followed by QA, and approved by a DER

S19

In considering a software development plan, ask questions: What tools are needed? Who will use them? What will the schedule look like? Which personnel does development, who does verification and who does QA? What environment will we work in? Which compiler sets? Am I developing in the UNIX system, Windows, etc.? What are the simulation tools? What are the debuggers?

You also develop a verification plan to verify the product using tools, processes and personnel. What is the schedule and time frame---how are resources allocated? You need a CM plan for configuration control of every data item. There must be a quality assurance plan to define how quality is inspected, based on your plans and processes.

Standards provide a consistent approach to requirement development, design and coding. DO-178B requires you to have these three standards written down. DO-254 has a similar set of standards and also asks for testing and production assurance through manufacturing.

That's an overall view to approaching the whole life cycle and to prove these plans are followed. As mentioned earlier, plans must address not just what you're going to do and when, but who is going to do them and how they get done.

They're typically written by the engineering team, not necessarily the DER or program manager. Plans need to be written by your team, while QA ensures not only that plans are followed but are acceptable, and comply with all aspects of DO-178B. The DER approves them to be sure they meet the intent of the document.

When you arrive at this point, you have a contractual basis; five plans and three standards. When they've been blessed by everyone, you can start building. But as idealistic as it sounds, it doesn't always happen that way. In seven out of ten programs, the PSAC is still being updated or negotiated on the day you're going to production. The reason is that reality often presents a different picture. Software can change during development, or your customer asks for features which require additional tool sets. These changes may not be compatible with the great amount of detail you're bounded to. The original plans are often done in the first three or four months of the program.

Let's say the program is three years long. You have to specify enough detail for approval, but still have sufficient flexibility to comply with the plan. For example, you choose a compiler and decide, "I am going to use company X's software test tool." If you specify an exact version at the time you use it, you could be locked prematurely into an approval. Over a period of time other versions evolve, so you may choose the newest release, say Version 7. When you wrote the plan you used Version 2.

Thus, you need to be flexible. You may want a version to capture structural coverage of the code based on testing. But you don't want to get into exactly how you're going to set up that tool in every detail in the plan because the environment may change.

Your plans, for example, might run 20 to 30 pages each. If you write a 150-page plan, you've gone too far and are into procedures. So develop plans with the "what" and a little "how" in them, but not with such detail that you're locked into something you may not want later.

**Evolution of DO-178**

| Document | Basis | Themes |
|---|---|---|
| DO-178 | 498 & 2167A | Artifacts, documents, traceability, management and contracts and testing |
| DO-178A | DO-178 | Processes, testing, components, four criticality levels, reviews, waterfall methodology |
| DO-178B | DO-178A | Integration, development methods, data (not documents), tools, COTS. |
| DO-178C | DO-178B | Reducing subjectivity, address modeling and reverse engineering |

s28

Soon after DO-178 was published around 1980 it raised questions: How should I conduct a Preliminary Design Review (PDR)? What is the format for documents? What is software's reliability number? The probability of failure is 1---it will fail---because I cannot fully test. An alternative means of compliance is proving to the FAA that you can test software 100% for every input combination, and verify every possible output. You just used "alternative" means of compliance," one of the few "loopholes" within DO-178B. But is it really a loophole? No, it's merely a well-justified replacement step for one of the steps required by DO-178B. Can you use "alternative means of compliance" often? No; only when you're truly stuck and you need to replace one of the required DO-178B steps by something that will work for you; it has to be at least as sufficient as that step within DO-178B which you're replacing. It's mostly common sense, and written justification. Just make sure your QA person and

DER formally approve of your "alternative," in writing, before taking the alternative.

It can be done with certain pieces of software, for example, a simple sin function; with a graph you can test at every point and boundary point and, therefore, does not require the other DO-178B tests. In using this alternative means of compliance every possible input and output are verified. But can you do it with 10,000 lines of code for, say, flight control surfaces? No, because there are too many variables and input conditions.

So, you like this "alternative means of compliance" and want to avoid explicitly testing your operating system by simply showing that you exhaustively use it. But wait, why can't you test the operating system that way? Because what is the combination of inputs into an operating system and the combination of outputs? You don't know it's a large number that is simply uncalculable and, therefore, unknown. So you cannot apply exhaustive testing to such an "alternative," therefore, that alternative is not valid. So it was decided that software components were complex and cannot be fully tested. What we need is a design assurance process. You develop the product and assure that the level of software integrity is the same as for complex hardware. If a resistor can be tested completely, for example, you no longer need to follow DO-254 simply because you can test every possible input and output. That means the hardware is "simple;" otherwise it is "complex" and must be certified via DO-254. That's all that's needed.

But can that complete and exhaustive testing, such as what you can actually do for a resistor, be done with a million-gate array ASIC? The number of test cases becomes 2 to the power of a million and you cannot possibly run all test cases. Therefore, follow the design assurance process which DO-254 or 178B provides.

When DO-178A first appeared it was considered very good. That was a time, however, when the software industry was progressing rapidly with new tools, methodology and languages. DO-178A could not deal with all the leaps and bounds so DO-178B was developed to address structured methods, tools, asynchronous real-world end-item testing and alternative methods.

We recall working on a new airliner, knowing that DO-178B would be published in the early 1990's. An application for a type certificate was submitted early because our client wanted to address only DO-178A as the certification basis. Today, there is a rush of applications to certify aircraft under DO-178B, considering that a new DO-178C may be more restrictive.

DO-178B realized that you don't really need just a document or follow the old Data Item Description (DID). You don't have to follow Mil-Std 498's processes. You just need available data and a process model that captures your requirements and design. That's good enough so long as you can keep it under configuration control.

A configuration management plan was formerly required. but now process and data describe it. That data may be combined in a software development plan with no further need for a document---just the data. The FAA does allow leeway on development of data items.

Artifacts show that you reviewed your software. You may be asked, "What's the artifact that proves you did the review?" It's like the use of receipts to prove deductions on income taxes. When an accountant says, "You want to take deductions for four briefcases this year---okay---but give me the receipts."

Checklists are artifacts, like footprints left as you walk through snow. Some people call them "breadcrumbs"----or what you leave behind throughout the life cycle process proving what you did. They're normally stored in paper form in cabinets or as electronic files in software development folders or CM.

We recommend the use of CM tools. CM tools are not formally required by DO-178B, but it's almost impossible to prove compliance without a CM tool. Also, such tools greatly improve efficiency. They enable you to check and retain everything and organize revisions, which means software development plan Version 1 is the one we like. A review of Version 1 exists in this artifact. That document and revision number are organized (we do it by company). Are you willing to publish how your company processes are established and that you've met SEI CMMI Level 4? Probably not because competitors can use it to avoid the $5 million you spent to get there.

Compare it to buying a car. You want it to go from zero to 60 mph in three seconds, get 50 miles per gallon and carry a family of seven. You also want it to look good, and not cost over $15,000. But what you probably get in the real world are two out of five. Similarly, each company has different goals. For some it's schedule, for others it's limited resources, so different processes are chosen to accomplish similar tasks. There is no singularly perfect tool.

When several DERs look at the same project---and they all disagree---all are correct! It's an interesting situation, and you need to manage it. In avionics and aviation today, the largest variable is the DER. He may make you do something that costs money, causing you to look for another DER who is more reasonable, flexible and easier to work with. He also may cost less and save time. That doesn't mean he's "pencil-whipping" (making false entries for work not done). In his mind, he feels you've met compliance which, incidentally, is what all DERs are supposed to do. They don't make regulations or say how the process should work. Their job is to determine if you comply with DO-178B or 254.

## Five Key Plans

| 1. PSAC | 2. QA | 3. CM | 4. SWDP | 5. SWVP |
|---|---|---|---|---|

**PSAC** Plan for Software Aspects of Certification

**QA** Quality Assurance Plan

**CM** Configuration Management Plan

**SWDP** Software Development Plan

**SWVP** Software Verification Plan

** *Plus 3 Standards: Requirements, Design and Coding*

S27

It's little different for DO-254. The plans are the same except to remove the software and insert hardware. Quality Assurance was changed to "Production Assurance" because it's the same set of steps and data, with concern extended all the way through manufacturing the hardware. These plans become the contractual basis with your team and the FAA as the blueprint of how you will build this product and the approach to process assurance.

## Plan for Software Aspect of Certification (PSAC)

| 1. PSAC | 2. QA | 3. CM | 4. SWDP | 5. SWVP |
|---|---|---|---|---|

20-30 pages

Overview of System, Architecture, & Software

Discusses Safety, Criticality, & Certification

References to other plans & artifacts

Help prepared by and approved by DER

Submitted to and approved by FAA

S28

It's not necessary or even desirable to write a 200-page PSAC, so keep it under 30 or 40 pages. If it's too detailed, it creates too many areas for questions at too early a stage in the program.

**Overview of system, architecture and software** must be within the PSAC. Why is it important to have a system description here? Because software is not certified on its own but as part of a system. That's why you need a system overview that shows architecture.

You must know whether the software has redundancy and monitoring built in. What language will be used? What's the microprocessor? Will there be partitioned software, Level A and Level C, running on the same microprocessor? How will their integrity be shown?

**Discuss safety, criticality level and certification.** You have to define criticality level in the PSAC; whether it's Level A, B, C, D or E and justify why. That could be a pointer back to data items done previously, such as system safety analysis, but you need to justify why you are at the chosen level.

**References to other plans and artifacts**. After initiating quality assurance, you write the approximate 30 pages of the QA plan. There is an approach for it, and the QA plan details exactly how quality will be assured in compliance with DO-178B. Get help in its preparation and have a DER at least review it, if not give approval.

The FAA almost always reserves the right to approve the PSAC. They like a recommendation from a DER but the approval is in their hands. There are many cases where officials simply say, "Charlie, can you approve this for me? Send me the final form that says you approve it." ("Charlie" is a reference to Charles Soderstrom, a highly experienced DER.) This "delegated authority" is becoming more frequent in some areas due to the FAA's increased workload---and you need these plans to be blessed.

**Quality Assurance (QA)**

| | | | | |
|---|---|---|---|---|
| 1. PSAC | **2. QA** | 3. CM | 4. SWDP | 5. SWVP |

- Addresses role of QA throughout process
- Ensures that all plans are coordinated and integral part of process, and are followed
- Ensures that transition criteria are adhered to
- Addresses conformity reviews and inspections
- Provides guidance and timelines for audits/reviews by QA (including checklists)

S29

The QA plan addresses quality or process assurance throughout the life cycle. What will the FAA do in this program? Who is it going to look at it? How often does QA conduct audits? Will they witness testing or code reviews? Are they going to participate in some processes and activities spelled out in the plans? Who do they report to?

Well, my QA reports to me. I'm responsible for delivering the widget and if he's not a good QA, which means he doesn't like what I do, I'm going to find somebody who does. That's not good. FAA will say that's not a good process. He should report to my boss who's responsible for quality of the company. And remember, QA cannot author any of the actual engineering activities (requirements, design, code, tests). QA's primary role is to ensure that your five key plans and three standards fully convey the applicable DO-178B criteria for *your* project, and audits are performed to ensure engineering is following those plans and standards.

He can report to me on a lot of his activities but I can't fire him because I didn't like the way he picked on my development, saying, "Your engineers didn't do this or that." That's why there must be a different organizational reporting process, which can be a strange thing. In the plan for building a home, you don't want plumbing and wiring that are not integrated, or the fuse box located underneath the hot water tank. You don't want a CM plan to conflict with a development plan; they need to work as a team or "system".

QA ensures those activities, for example; transition criteria. This means going from one process of software development to another. You need entry and exit criteria before moving to the next process.

In our house-building analogy the first step is pouring a foundation after all plans are approved but don't start framing until the city inspector says, "You poured the right amount for the foundation, it's the correct strength and material. Go ahead. Approved." You met the entry and exit criteria.

If framing is started on top of a foundation that's not approved because entry and exit criteria weren't met, I may have to tear it all down. In software, you need a good set of requirements, then follow the process of reviewing, accepting and making all iterations before going into design and coding. In earlier days, we had PDR/CDR. Those are, in reality, transition criteria and were emphasized in DO-178A.

The subsequent DO-178B recognized that products aren't developed that way, but are iterative. Take just one requirement and go through a transition to design because the requirement document isn't finished overnight. It may take 15, 20 or 30 PDRs so it's done incrementally and called transition criteria.

Be certain that development conforms to plans and production of the manufacturing processes. Put in the QA plan when you're going to do what, how often and why you're going to do more if you find failures.

We encountered one FAA person who insisted on percentages in the QA plan. What percentage of code will be audited by QA? The plan said that QA will audit the source code and review process, but there was no percentage. The FAA desired 10, 15, 20 or some other number. DO-178B doesn't give such guidance but, in this case, we put it in and it was approved.

Let's assume the QA people say, "10." And FAA replies,
"Not good enough."

"Well, what do you want?"
"One hundred percent," says FAA.
The QA guy says, "Not good enough. That's not an audit anymore. That's a complete review."

Now the DER gets involved and a negotiation results in, say, 40%. Now, everybody is happy. But after much arguing and bargaining a percentage was agreed upon.

QA is a process so there must be a method to assure the FAA that its work follows the process. Remember; QA's primary role is to ensure all project processes, plans and standards comply with DO-178B, and that everyone follows them and can prove it. That's why it needs checklists, breadcrumbs and artifacts. FAA relies on QA, as does the DER. The DER ensures that the QA, developer and verifier did their work and found as many errors as possible in the development phase before transitioning. The FAA ensures that everybody has done the right thing for compliance. Normally, problems get caught between these levels of audits.

In the CM plan there must be tight configuration control of the product, data item documents and artifacts of the lifecycle. It will also define a naming convention and numbering with the revision methodology.

The historical CM system (still used by some companies) is similar to checking out books at the library. The librarian stamps a card and, two weeks later, the book is returned. That system of configuration---checking in and out---has mostly been replaced by automated systems. They're based in your network, so you define the system and how you and the developers are going to use it. Criteria are established on how it's set up, who has access, who doesn't, reporting problems and how the system is tracked. Changes in software may not be made without giving reasons; what failed, how it was fixed and an approval record.

Security is also a big concern today. How is the system backed? How would you rebuild the software, and who is responsible for CM? The answer to the latter is everybody who works on the system needs to know how to write a problem report and how to check data in and out. These documents must be available to all engineers working on the program.

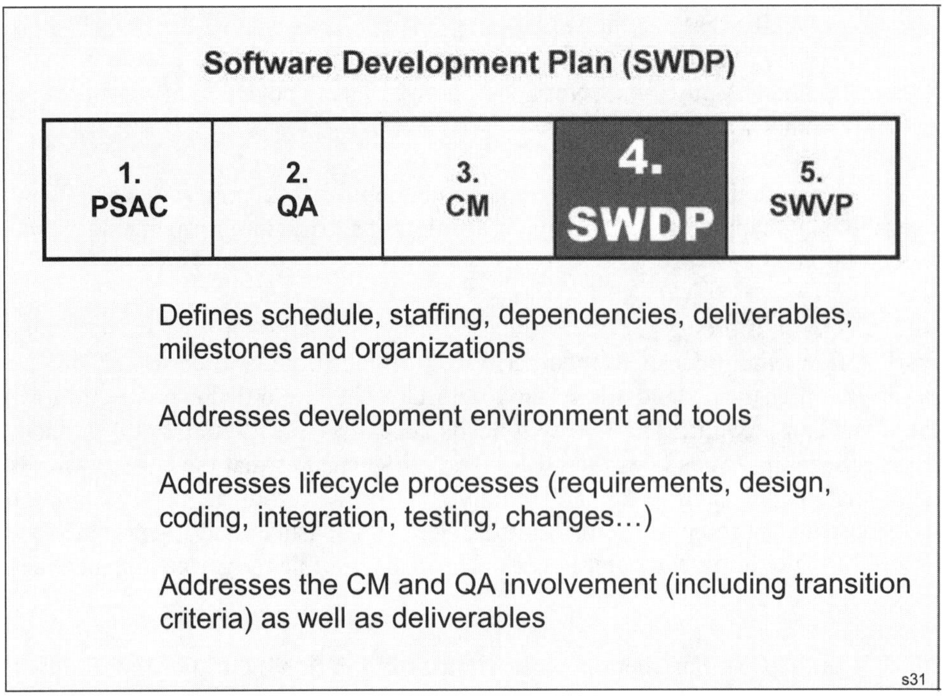


**The Software Development Plan** is for staffing schedules, deliverables, milestones, tools and the software development environment in which the software is designed, developed, implemented, then integrated. It addresses life cycle processes, how requirements are defined and whether you're doing model-based development. Will you use structured methods? Are you utilizing ADA or C? What are the tool sets and your requirement traceability?

Are you using DOORS to track your requirements, Engenuity's QCG to automatically generate your graphics software or Integrity-178B as your avionics RTOS? Those details go into the development plan. How are you going to set up and use it? And you want to address unit testing if you do static analysis. The environment for development may be different than your target. For instance, you may use PC Windows or a UNIX system. They need to be defined here.

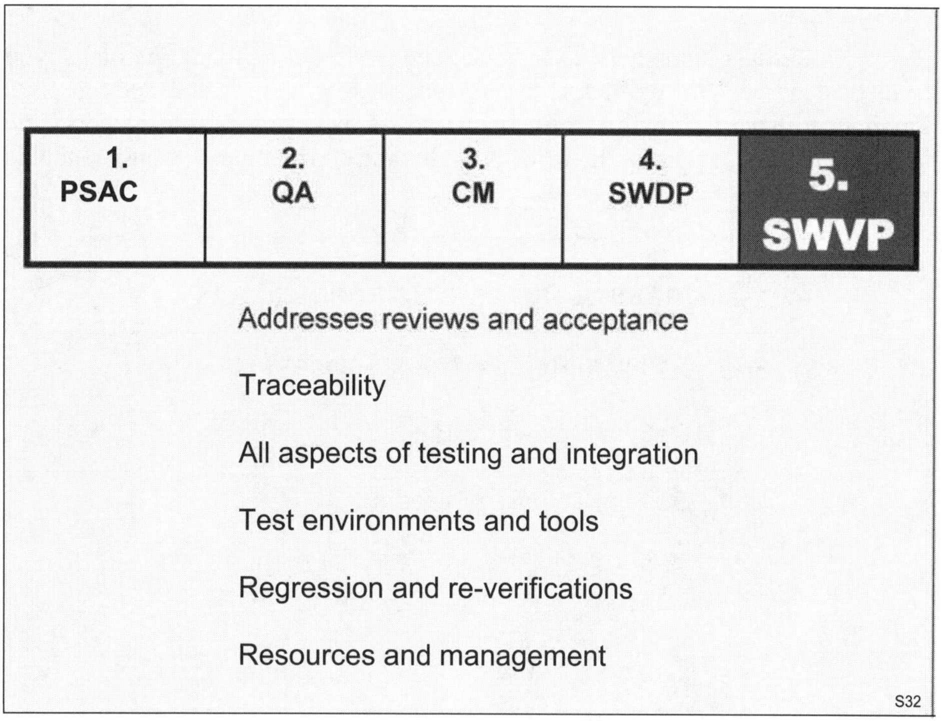


**The Software Verification Plan is** for staffing schedules, deliverables, milestones, tools and the software verification environment in which the software is reviewed, verified, then proven correct. It addresses life cycle processes, how requirements, design and code are verified. Will you use formal tools? Are you utilizing tools such as Polyspace to check your code prior to peer review? What are the tool sets and your requirement traceability?

Are you using VectorCast to test and structurally analyze your code coverage? Those details would go into the verification plan. How you are going to set up and use the tools? You want to address where you'll test your hardware/software integration tests and the regression testing strategy.

Other data items need to be generated. After completing the plans, you develop standards for writing requirements, writing design and coding. It's important to have thorough standards because if there are ten developers, you don't want each to bring their own style and formatting. Consistency is important.

**Additional Documents and Artifacts**

Software Requirements Standard

Software Design Standard

Software Coding Standard

Software Configuration Index (SCI) or Version Description Document (VDD)

Traceability Matrix

Requirements, Design, Code & Tests/Results

Tool Qualification Plan/Data/Assessment

Software Environment Configuration Index (SECI)

Software Accomplishment Summary (SAS)

CM Records & Problem Reports

QA & DER Audit Records

S33

It's also important to address safe design and coding concepts. For example, C++ has characteristics you don't want your coders to implement so define an acceptable subset of C++ and specify that in your coding standards. An entire program can be written in one line but that's not good coding practice.

How many characters in a line, how many functions in one file? One file should not contain every function, so define in the coding standards some basics and more complex issues such as maximum allowable McCabe value. Recurring C code creates problems. There is stack overflow, which isn't safe. Can you use inheritance within C++? No. For Level A? Also, no. You want to bound those activities or be able to analyze them, so standards need definition.

**Configuration index or version description document**. VDD is old terminology, but still used by the military. It's the same as configuration index, which contains a few additional pieces the VDD never called for, but needed for the configuration index. That's not just for the product but the environment in which the product is developed.

Next, there is a **traceability matrix.** Requirements, design, code and test results are actual results of activities you perform. Software environment configuration is the environment in which you develop the product, test it and the tools being used. The trace is top-to-bottom and bottom-to-top, as previously explained.

**Accomplishment summary** is what you did. The PSAC says what you want to accomplish but if not followed exactly, state where and why you deviated and the justification.

**Records of quality and problem reports** are generated and addressed throughout the life cycle, including records of where the DER was involved, checklists generated for each process step, audit records, and artifacts throughout the life cycle.

**Avionics system documents** have specifications that may come from your customer to guide in developing the product. The customer might also provide the system design and ask you to develop it. A system design may include multiple LRUs within the same system, redundancy and different paths. So the number of units that do the same thing is part of system design. Once established, system design divides into two areas. The functionality will perform either in software or hardware requirements.

Before arriving at this point, however, you must go through a development process for system architecture and design through a safety analysis that looks at functions, failure and safety information. This detects such unsafe conditions as a single point of failure which causes the whole system to fail. For example, if all power to modules that control flaps on an aircraft comes from the right engine, failure in that engine causes loss of flaps, as well. To avoid this, you want multiple power sources from different power generating systems.

A hydraulic system might have five redundant power sources; with four electronically controlled. The fifth on some aircraft is called RAT, for Ram Air Turbine; if all power fails, a door on the belly of the airplane opens and out drops a little fan that spins in the air stream to generate emergency power. It's a simple, but last line of defense.

In looking at design assurance processes, note that the word "firmware" no longer appears. There is either hardware (DO-254) or software (DO-178B).

The former "firmware" refers to silicon-based logic, which is now solely called "complex hardware".

DO-178 and DO-254 deal with systems and DO-254 normally addresses ASIC, PLD and FPGA complex electronic parts. DO-178B also deals with the board support package (BSP), real-time operating system (RTOS), the actual application software, areas of math libraries, drivers and graphical drivers. As seen in the figure below, it breaks up the LRU into how parts are controlled and categorized.

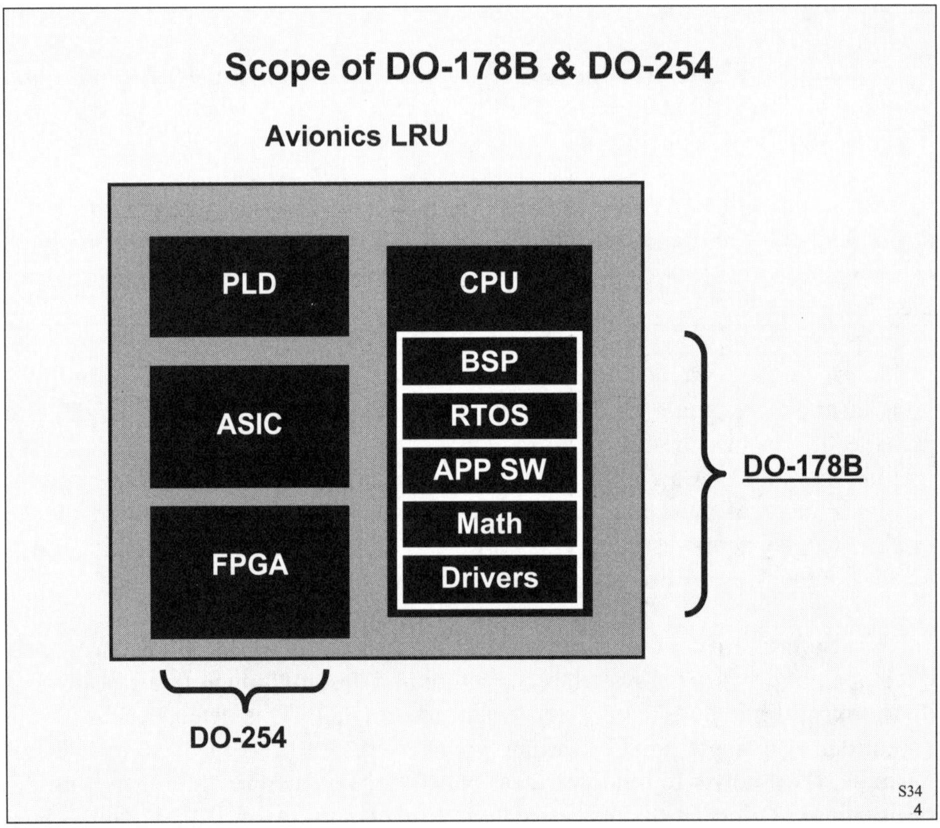


And that is how you plan your project; by applying the aforementioned considerations and strategies. Now you're ready to apply DO-178B and DO-254 and where to start.

# 4. Criticality Levels

SAE standard ARP4761 on conducting a safety assessment. determines criticality level----A, B, C or D. Sometimes your customer tells you the level. Is his conclusion open to a meaningful, friendly dialogue about lowering the criticality? Probably not.

Criticality levels are defined in the document but, informally speaking, Level A is the most stringent, where a failure causes some or all passengers to die, followed by failure at Level B, where some people may die, then Level C, where no one dies but there are major injuries. At Level D there are minor injuries and Level E has no impact on flight safety.

If you're at Level E, you don't follow DO-178B. You take the Level E designation and do not have to apply DO-178B or DO-254 to your avionics system. Can simple items like lavatory lights or passenger entertainment systems be Level E? Not always. For example, bathroom lights may connect to independent power sources so they can never be turned off. If power is lost from one source, there is a battery on a hot backup. Why go to that redundancy? Let's say you're on the ground and smoke fills the cabin. The crew calls for evacuation of the airplane; you don't want to be stuck in a dark lavatory looking for the door release. Thus, even lowly lav lighting deserves Level D certification.

How about passenger entertainment, say, a seat-back video screen? It seemed unimportant until a real-life system malfunctioned and propagated smoke throughout an airplane in flight. On reaching the flight deck, smoke triggered events ultimately leading to a fatal crash. Investigators later found the entertainment system had shared a bus that delivered primary power to critical cockpit systems. So criticality levels can be very important, even between apparently unrelated systems.

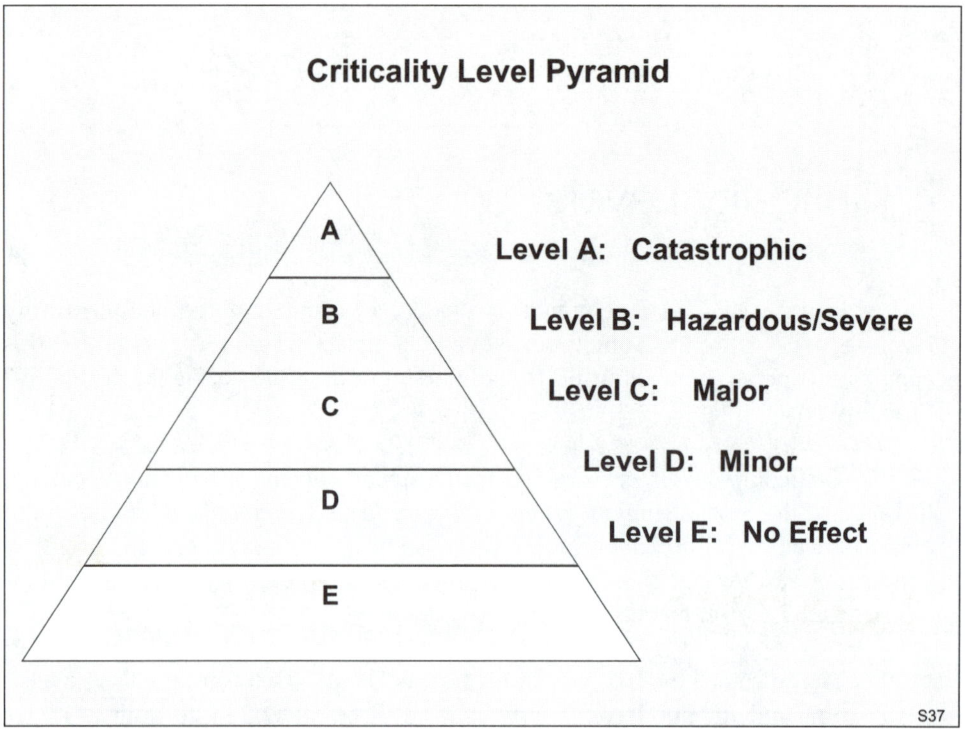


**Level A, Catastrophic**, often resulting in total loss of plane and passengers, can happen simply by putting too much workload on the crew. Some function breaks and suddenly the pilots are so preoccupied dealing with the problem they have a catastrophic event. It doesn't have to be a system failure or faulty indicator that denies some piece of data. It could simply cause so much work that pilots can't keep up with flying the airplane.

**Level B, Hazardous/Severe**, may not kill everybody, but there is potential for that hazard.

**Level C, Major**, can cause people to become injured and the aircraft uncontrollable.

**Level D, Minor**, may have some effect but it can be overcome by the aircraft, and pilots can maintain control.

**Level E**, has no effect. The system safety assessment process determines these criticality levels. If it's Level E, something establishes that level and you need to justify it. You can't just go to the FAA and say, "It's Level E. I don't have to do anything." If your customer didn't justify it, you will have to.

Typically, while designing a product or seeking a TSO you need that justification. DO-178B Appendix A describes what's needed for each level and that's what people study most of all. You may find shortcuts there but we guarantee you will spend most of your time arguing and justifying the shortcut, rather than following the process. It'll cost less and be done faster.

## Criticality Level Comparison

| DO178B Aspect | Level A | Level B | Level C | Level D |
|---|---|---|---|---|
| Independence Level | High | Medium | Low | Very Low |
| Necessity of Low Level Requirements | Yes | Yes | Yes | No |
| Statement Structural Coverage | Yes | Yes | Yes | No |
| Decision/Condition Structural Coverage | Yes | Yes | No | No |
| MCDC Structural Coverage | Yes | No | No | No |
| Configuration Management | Tight | Tight | Medium | Low |
| Source to Binary Correlation | Yes | No | No | No |
| Requirements Correlate to Target processor | Yes | Yes | No | No |
| Architecture & Algorithms Verification | Yes | Yes | Yes | No |
| Code Reviews | Yes | Yes | Yes | No |
| SQA Transition Criteria | Yes | Yes | Yes | No |

S38

**Statement structural coverage.** This is required at A, B and C.

**Decision tradition coverage**. Not required at Level C, only at B and A. **MCDC** is only required at A. This is a unique requirement of 178B as a metric process. It's not a testing, but an analysis process. Many people think they've got to test, test, test to achieve structural coverage. But it's not just coverage and it doesn't have to be achieved by testing. You can do it completely by analysis when there is no other ready alternative.

**Configuration management** goes from low to high as the criticality level increases. Most DERs (and FAA people) basically say, "As we increase the criticality level, we want more rigor."

"Rigor," however, is a vague word. DO- 178B doesn't indicate major differences in configuration management between A, B and C, but as you go from C to B to A, more rigor is required. In other words, problem-reporting, configuration management and status process have to be more accurate. It requires more artifacts and a higher level of description and change management..

We've never seen anybody write a different process for configuration management for these levels. But the DER will be looking for---if you're doing Level A---a better execution of that process. While doing Level C, there is a tendency to let things through.

**Source to binary correlation** is only done at Level A. FAA issued a paper for aircraft builders to answer certain questions, for example; if a compiler is used, a lot of people want to test on a host system using a host-based compiler, thus cross-compiling. Also, software developers also want to optimize the execution of their code by shrinking its size or increasing its execution speed. That is done via compile-time options called "optimization". But can you test, or qualify, a compiler? No, the job is simply too large to reasonably tackle and you'd be kidding yourself if you thought you could thoroughly test the compiler anyway. Besides, the output of the compiler undergoes testing, so the compiler is implicitly verified. But for a Level A system, you really must trust the compiler. So for Level A there is an added degree of rigor required and a step is added: "source to binary correlation". You must explicitly verify that your compiler emitted binary correlates with your source code, as intended.

If evidence testing is done on that host is that actually the system that will fly? No, it is not because final "for credit" testing must be done using the end-item target binary executable. To get credit for testing, you have to show a comparison of that binary to that source. It's best to use the same compiler all the time on the target. For Level A, test on the target so the source is represented correctly in the binary. In other words, you can write source but the compiler puts data into binary. There needs to be a correlation that the binary is correct, as opposed to just being a source.

A source code generator tool needs to be qualified because it automates DO-178B steps, so you physically don't have to perform them on every program. Once the tool is qualified it does the job because it knows how the code is generated and compiled, and how the object code comes out. There are four or five systems that do code generation and check their process, and they've added source binary comparison.

The FAA doesn't certify product tools. As you go to higher levels of criticality, there is more need for proof and higher workload. Code reviews are required from C and above, but not for Level D.

**SQA transition criteria**. Many standard QA code reviews and architecture requirements start at Level C. Most of the work is during the structural coverage analysis going from C to B to A.

Deciding on the criticality of software and hardware is based on a top-level process. That's how the PSAC and PHAC, for example, state you are at Level A. If I'm at Level B, I don't have to justify myself anymore but will point to the process that did it. Your customer may do that or you just may be told you're Level B and that's it because your customer's safety analysis supports that Level B designation.

Remember, at the beginning of the day, and at the end of the day, the Criticality Level should be determined by following a thorough and meticulous Safety Analysis process. That determines the contribution to safety of the overall system and the components within, and drives both the criticality level determination and basic system architecture.

# 5. What is "Certified"?

The FAA doesn't certify *software*, they certify *systems*---and can do it several ways; there's the TSOA, Technical Standard Order Authorization, and the STC, Supplemental Type Certificate. For new aircraft there's the TC (Type Certificate) and the ATC, for Amended Type Certificate.

> **Technical Standard Order (TSO) Versus**
> **Supplemental Type Certification (STC)**
>
> **TSOA**: Technical Standard Order Authorization– the development of a product such as GPWS, Radio, Weather Radar:
> Approves aircraft "appliances" for use on different aircraft
> Dual design and production approval
> Is not an approval for part installation
>
> **STC:** Supplemental Type Certification- the approval of type design changes not requiring a new Type Certification (TC):
> For major alterations to aircraft
> Available to non-TC holders and TC holders
> Includes part installation

Let's look more closely at the role of certification. In the real world, no one builds specific radios for every aircraft in the world. A navigation radio is designed, for example, that meets a minimum standard, then is certified under a TSO for that product category. But is the product now airworthy to fly on an airplane? In many instances---such as flight control and displays---there is another certification process, the STC, or Supplemental Type Certificate. This is for equipment added to the airplane after it's been manufactured and is in service. If you're the original aircraft manufacturer, it's called a TC (Type Certificate), which certifies the system on how it's installed and functions on the aircraft.

For transport aircraft, the rules are stated under Part 25.1301 and 1309. For smaller aircraft, it's 23.1301, 1309. For helicopters, it's 27 and 29.1301, 1309. Those regulations say the system must meet its intended function, which is the reason for DO-178B and DO-254. To help the industry comply, FAA developed guidance in the form of an Advisory Circular (AC) that recognizes RTCA document DO-178B as a means of showing compliance and that the software meets its intended function.

> **Terminology**
>
> **"Certified"**: an entire system is Certified, and components may have different certification levels
>
> **"Certifiable"**: a component within a system achieving its highest certification status prior to certifying it with the system ("Certified")
>
> **"Qualified"**: certification of a tool which, since it does not itself fly, does not require "certification"
>
> **"Compliant"**: certification via a different entity than FAA, e.g. military or non-commercial avionics.
>
> S39

An operating system such as Integrity-178B can be certifiable, but tools, other components, or board support packages are not, as they can only be verified when integrated with other software components. According to DO-178B, tools are "qualified," not certified.

Do you qualify a signal generator? Probably not. You qualify verification and design tools specific to the development process. If a tool in a lab generates a test pattern, there's no need to qualify that. You don't qualify CM tools because their output is verified via subsequent DO-178B steps.

DOORS, Synergy or whatever you're using, do not need to be qualified because they don't have anything to do with actual verification or development, and their outputs also are verified via subsequent DO-178B steps.

"Compliant" is a new kind of term, especially since the military has become involved and now almost always requires some form of DO-178B (and very soon, complete DO-254) compliance. The military has a tendency not to certify anything; it does what it wants and gets waivers. But "compliant" basically means you can meet the 178B specification because you've done sufficient work to show compliance. What's the difference when a DER looks for compliance? It's exactly the same job---no difference at all---only he'll never sign a form (such as an 8110-3) that says you're certified because that form is valid only with the FAA. Most likely, the DER or somebody experienced in certification will say you've done sufficient work to show compliance with 254 or 178B. However, it's not used in the FAA world.

# 6. Costs Vs. Benefits

Is DO-178B really expensive? Does it add value? Will it improve safety and reliability? What are the true costs versus benefits? These important questions deserve clear, straightforward answers.

We said that 178B and 254 shouldn't add more than 20 to 40% to the cost. However, you want to look at that cost, schedule-wise, because it eventually translates back to budget. From our experience over the years, industry shows increased cost at about 75 to 150%---and we've seen it reach 200%. Why such a disparity?

There are several reasons. Maybe your practices aren't well defined, or too much time is spent spinning wheels and iterating, or somebody came along and said "You must do this." Several DERs might say, "Thou shalt do it my way and *only* my way." If you don't comply, you end up doing a lot more work.

Some DERs still believe you need to do unit testing, although there is no requirement for it. That's a difference between 178A and 178B; unit (component) testing was downgraded between the two documents. Some people still have it in their process definition standards and it certainly adds value. Only if you define unit testing in your process will you need to do it. But DO-178B does not require it. Remember, DO-178B is about the "system," and verification is best when done on the system, or at least on integrated software that is exercised and stimulated as though it were operational. Unit testing is very subjective, and DO-178B emphasizes determinism, not subjectivity.

Typically, the unit usually is a 2167-type definition, single entry/single output, like a cohesive piece. Some people say, "No, we're going to call a unit large components, classes, larger classes, combination of classes, programs." And that should be part of your standard. If unit testing is done, you should define a unit. Always tie the units to functions and requirements of software.

We worked on a new weather radar for the C-17. The entire job for two years was structural coverage analysis. All we did was verification work to make sure it happened correctly. We did not do a single piece of unit testing. Had we done it, you would have seen how much code there was; 14 DSPs chained and two power PCs working together. It was a huge amount of code and number of units, as defined. We'd have spent another two years unit testing, which is costly because it is manually intensive. But if it's in your process, you need to follow it.

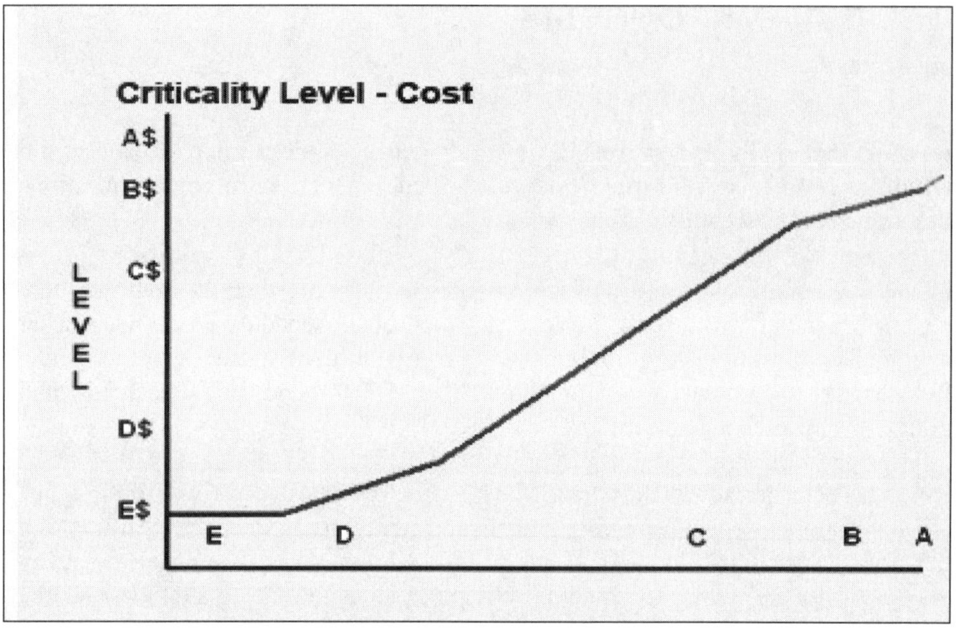


**Delta cost of DO-178B per criticality level.**

## Cost vs. Criticality

Consider the impact on cost by different levels of criticality, ranging from Level A (least critical). As the level increases, so does the degree of rigor associated with documentation, design, reviews, implementation and verification. Accordingly, cost and schedule increase, as well. The figure above and table below depict actual cost deltas associated with each level of criticality.

If you look at the schedule going across time and see the cost of A, B and C, D, the major jump is between D and C. As criticalities go up, we're representing that cost is actually less. There's little difference between A and B. The only large item that's different between A and B is MCDC.

| Level E Cost | Level D Cost | Level C Cost | Level B Cost | Level A Cost |
|---|---|---|---|---|
| Baseline | E+5% | D+30% | C+15% | B+5% |

Next, the myth that DO-178B is expensive. As seen above, Level D certified software still has full planning, implementation, reviews and basic testing. Configuration management, quality assurance and DER liaison are applied to Level D, but the costs are hardly more than any non-certified software process. It's because Level D is comprised mostly of normal, "good" software engineering principles.

Another myth is that the most significant cost escalation occurs when moving from Level B to Level A. That is untrue; the cost impact of DO-178B is most significant between Level D and Level C. Why? Level C requires the following, which Level D does not, and results in Level C requiring 30% more budget and schedule:

Testing low-level software requirements
Ensuring 100% coverage of all source code statements
Greater rigor placed upon reviews
In many cases, more rigorous configuration management

Level B requires additional structural coverage (decision-condition, e.g., all branches in the source code), additional independence in reviews and tighter configuration management.

On first glance it seems that Level B should be significantly more expensive, e.g. 50%-70%, than Level C. In theory it seems to make sense. But as in many areas of life, common sense overcomes theory. In Level B (and C), there are detailed low-level software requirements and they must be thoroughly tested. During such requirements-based testing, the great majority (70-90%) of branches in the source code are already executed and require no additional structural coverage testing if test capture and coverage tools are appropriately used! Thus, the seemingly significant cost increase associated with Level B versus C structural coverage is already mitigated by requirements-based testing. And the added structural coverage requirements are largely automated using pre-qualified tools such as VectorCast, which greatly improve efficiency. Quality software engineering organizations already incorporate a semi-automated and streamlined process which includes independent review and tight configuration management. Thus the added cost of those aspects for Level B is lessened. The reader is well-advised to undergo DO-178B training and DO-178B Process Improvement to leverage these cost-reduction techniques.

Another misunderstanding: Level A is the most critical software level and the most expensive. But another myth exists for Level A, namely ***"Level A is very difficult to achieve and costs at least 30-50% more than Level B".*** That assumption is false. Level A imposes yet more structural coverage requirements (MCDC testing), robustness and correlation requirements, and slightly more stringent reviews. The most significant cost driver over Level B is the MCDC testing requirement. However, with proper application of modern structural coverage tools, personnel training and thorough requirements-based testing, the added cost for Level A can be largely contained.

DO-178B is not free, but can be cost-effective when implemented properly. Why, then, are so many entities adopting DO-178B? The following are our conclusions about the value of DO-178B, based on actual experience:

**Greater upfront requirements clarity**. DO-178B mandates thorough and detailed software requirements. Such detail and the necessary discipline force answers up-front instead of being deferred. Assumptions are drastically minimized. Consistency of requirements and their testability is assured. Iterations and rework due to faulty and missing requirements are greatly reduced using verified requirements traceability.

**Fewer coding iterations**. Code iterations, or churn, are the bane of software engineering. In many cases, 10, 20 and even 30 versions of evolving code files exist on new products. That's nonsense because code should be largely correct the first time it is written and not require dozens of updates to get it right. Code should be reviewed by analyzing implementation versus documented requirements. The use of solid code analyzers will further help to reduce code iterations.

**Fewer bugs during unit testing.** Since DO-178B mandates thorough testability requirements and code reviews, far fewer bugs are found during unit testing. And independent code reviews further that aim.

**Greater consistency within software**. Software is like a chain; only as strong as its weakest link. Software that is 99% correct, is 1% incorrect, which means unsafe. The weakest software module, or software engineer, is on the critical path of software safety. All software must be consistent per its level of criticality and DO-178B enforces this. Remember; DO-178B and DO-254 are about "determinism".

**Fewer defects during integration**. Integration can be a lengthy, iterative process where major defects requiring design changes are uncovered and fixed. Not with DO-178B, where integration is typically 50-75% faster than non-DO-178B environments.

**Improved configuration management**. The ability to control and recreate any snapshot of the project is covered by configuration management. But CM also ensures security, backups, completed reviews, problem reporting and version control. DO-178B's strong CM requirements, coupled with modern tools which automate many CM tasks, ensure higher software quality both now and in the future.

**More thorough testing.**

**Easier regression testing.** You build a product on the assumption that it will be successful and have a long life. And you know the software will evolve through new applications, installations and versions. Regression testing can be expensive if extensive or manual. DO-178B provides thorough traceability to determine which modules need changes or analysis and retesting.

**Improved hardware integration**. Integrating software onto its target software is typically challenging for embedded systems: the development environment is quite removed from the target environment and implemented via different engineering teams. DO-178B mandates that software components which could be affected by hardware be tested on the hardware, e.g. hardware/software interfaces, interrupts, timing, board-level components, BSP/RTOS, etc. And the improved determinism and quality of all these components via DO-178B belies improved hardware integration.

**Fewer in-the-field defects**. In-field defects and customer returns and 're-calls' are extremely expensive in the short and long-term. Products developed under DO-178B have been achieving 80%-90% fewer returns.

**Improved re-usability.** Thorough and consistent documentation required by DO-178B, modularization, enforcement of documented modern engineering principles and reviews to ensure all the above, greatly improve software re-usability (and, in software, re-usability is the Holy Grail). But the reality is that unless a software component is at least 80% re-usable (e.g. unchanged), it is quicker and less risky to simply start from scratch. And most software is less than 50% "re-usable". With DO-178B and enforcement of design/coding standards, coupled with independent reviews and traceability, most modules should be at least 90% re-usable.

**Improved customer satisfaction**. Reliable software begets reliably returning customers.

**Decreased single-point personnel failures**. Software is a technical art--- and artists resist documenting their work and subjecting it to common development standards and peer reviews. Without standards, discipline and modern software engineering principles, software teams devolve into a group of loosely structured rogue artists. They are highly valuable, creative and talented, but the loss or deficiency of any such artist is catastrophic to the team. All work must be documented, understood and consistently applied. DO-178B greatly reduces the possibility of such single-point personnel failures.

**Improved management awareness of true schedule status.** How many software projects report "99% complete" week after week? How is software progress measured? How can management truly determine the

| Benefits of DO-178B | |
|---|---|
| Greater up front requirements clarity testing | Fewer in-the-field defects |
| Fewer coding iterations | Improved customer satisfaction |
| Fewer bugs during unit testing | Improved re-usability |
| Greater consistency within software | Decreased single-point personnel failures |
| Fewer defects found during integration | Improved management awareness of true schedule status |
| Improved configuration management | Improved completion schedule due to the above |
| Easier regression testing | Wider market acceptance |
| More thorough testing | Improved differentiation *vis-a-vis* your competitors |
| Improved hardware integration | |
| **The Military Has Adopted DO-178B Because of:** | |
| Improved supplier consistency & visibility | Improved system reliability |
| Improved software quality | Lower lifecycle costs |
| Greater reusability | |

completion status of software? The answer is modern and accurately detailed management techniques built around DO-178B. It has provisions for insight, traceability, accurate status on design, development, testing, integration and reviews.

**Improved completion schedule due to the above.** "What gets measured gets done." DO-178B provides for continuous project completion and quality statusing.

**Wider market acceptance.** DO-178B provides widely-known and accepted proof of reliability. Safety-critical applications abound in aerospace, medical, transportation, energy and even entertainment (such as amusement parks). These fields put lives at stake and are in need of improving safety and reliability.

**Improved differentiation from your competitors.** Our economic system assures that competitors compete, which means that someone will adopt your product's best features, thereby eroding your differentiation. The answer? Evolve your own products and differentiate them yourself. Companies and products must continuously grow to survive. Prove reliability, prove quality and prove your value-added proposition: DO-178B, far from a rubber stamp, enables you to demonstrate such proof.

# 7. Military Certification

For decades, military organizations developed hardware and software using defense-oriented standards, including 2167A, 498 and 882. They were highly motivated to use standards which differed from the commercial sector since their applications were perceived as "different". To fulfill the military's utmost concern---"mission"---they applied the following rationale:

Military projects are more complex than commercial projects
Mission completion is always a highly desirable goal
Military projects need higher quality than civilian projects
Military projects have numerous suppliers to manage
Military projects require complex integration cycles
Military projects need long airframe lifetimes

Before the appearance of DO-178B, that rationale was valid. Today, however, there is much commonality between military and commercial avionics for a growing number of reasons:

Both utilize high complexity and complex integrations
Both draw on hundreds of suppliers with long project lifetimes
Both require access to leading-edge commercial technologies
Both are increasingly concerned with re-usability, quality and increased cost-effectiveness
Military aircraft are utilized increasingly in commercial airspace and do not want restricted flight paths or hours
Suppliers covering both military and commercial realms, and merging their product lines

By the year 2003, U.S. military organizations realized that commercial aerospace, particularly sectors regulated by the FAA via DO-178B, maintained certain advantages not available to the defense establishment. They faced these choices:

Maintain status quo and do nothing
Update Mil standards with the best aspects of DO-178B
Simply adopt DO-178B (and later, DO-254)

Their choice was the last one; simply adopt both documents.

It wasn't easy. As in any established organization, there were myriad opinions, entrenched practices and opposition, added transition costs and politics. The result? A gradual adoption of DO-178B and DO-254 (although DO-254 in the defense sector lags that of DO-178B). Also complicating military adoption of DO-178B were other considerations.

The military did not want to relinquish oversight to the FAA, nor did the FAA have the bandwidth or authorization to intervene in military projects. The military, too, was unfamiliar with DO-178B and DO254 specifics, and applied widely varying, subjective criteria.

Truth be told, the two documents *are* terse and vague. Specialized training by experts is typically required to apply them in the real world.

Because of these factors, the military may complicate DO-178B /DO-254 by continuing to require adherence to MIL standards as well as de-emphasizing key areas of DO-178B/DO-254. This is counterproductive since the standards differ and are often in conflict. Nevertheless, DO-178B and DO-254 are gaining rapid adoption by militaries worldwide and the authors predict it is here to stay.

## Compliance versus Certification

Since the FAA, with few exceptions, is not involved in military projects, formal certification is not required. Instead, military agencies typically self-certify under the term "*Compliance*". They require compliance to DO-178B/DO-254, not certification, and the FAA is not involved. DERs are not required (although they are advised) and certain requirements are relaxed, including some aspects of safety analysis, code robustness, source-to-binary correlation and strict MCDC proof of all instances.

## GAP Analysis

Most military services and suppliers establish generally high-quality, organizations and processes. When adopting DO-178B and DO-254 they can re-use much of their existing processes, documentation and artifacts. Often, they operate at 60-70% DO-178B and 30-50% DO-254 adherence without even considering DO-178B. Therefore, when faced with the requirement to "comply with DO-178B", it is most cost-effective to do a Gap Analysis. This assesses their current processes and determines the "gaps" compared to full DO-178B (or DO-254) adoption. A Gap Analysis typically takes 2 to 4 person-weeks by DO-178B/DO-254 experts, and can save years by maximizing re-use. Unlike military standards, which have strict requirements for document/artifact format/content, DO178/DO-254 provide greater latitude, therefore, retention, and re-use of existing items.

When we perform a DO-178B/DO-254 Gap Analysis for the military, the following levels of "Gaps" are typically found within the audited organizations (where "0% Gap = 100% Compliance):

SEI CMM Level 1 Organization: Gap = 70% – 90%
SEI CMM Level 2 Organization: Gap = 50% – 75%
SEI CMM Level 3 Organization: Gap = 35% – 60%
SEI CMM Level 4 Organization: Gap = 25% – 40%
SEI CMM Level 5 Organization: Gap = 20% – 35%

On a DO-178B/DO-254 individual activity basis, the particular Gaps are typically as shown in the following table:

| CERTIFICATION ACTIVITY | % "GAP" |
|---|---|
| • ***PSAC*** | **80%** |
| • ***QA Plan*** | **20 - 30%** |
| • ***CM Plan*** | **10 - 20%** |
| • ***Development Plan*** | **40 - 50%** |
| • ***Software Verification Plan*** | **60 - 70%** |
| • **Safety Assessment** | **80 - 90%** |
| • **Requirements Definition** | **20 – 30%** |
| • **Design** | **10 – 15%** |
| • **Code** | **5 - 10%** |
| • **Functional Test** | **5 – 10%** |
| • **Structural Coverage Tests** | **90 – 100%** |
| • **CM** | **10 – 30%** |
| • **QA** | **50%** |
| • **Tool Qualification** | **100%** |
| • **Checklists** | **30 – 50%** |
| • **Reviews** | **30 – 50%** |
| • **Audits** | **30 – 50%** |
| • **DER Liaison** | **100%** |

Based on some 200 military projects completed in recent years, the same benefits discussed in the previous chapter are achieved, such as greater supplier visibility, clarity of upfront requirements, fewer coding iterations, greater software consistency, fewer defects during integration, easier regression testing, fewer in-the-field defects, higher management awareness and improved completion schedules.

# 8. Getting Started

There are basically two ways to adopt 178 and 254; the big bang theory or evolution. If you're a newcomer we recommend the big bang. Avoid the SEI process of going through Level 1, 2, 3 and 4 because you have too large a gap to close via gradual evolution. Instead, you'll need to "kick the habit" of informal development processes cold turkey; just bite off the whole 178B and incorporate it into your program. Don't approach it in pieces---DO-178B isn't a piecemeal entity so it doesn't work that way. Literally, the sum of DO-178B is as large as the whole. And you won't confuse the developers as you would if you evolve deficient processes to comply with DO-178B. If you try to fill a large gap piecemeal, the problem is that you train people to do one piece---and that's the only piece they'll do. A year later you add another piece of the DO-17B life cycle process and it's different. Now there's confusion because of too many disjointed processes.

Just write one process for the five different software life cycle planning processes and get it in place. Train your people to develop the whole system via DO-178B. Finalize the five key process plans after performing your gap analysis, then identify third party tool sets. Plan the tool qualification and coordinate with the FAA certification agency. It's easier to plan all the pieces at one time. Remember, DO-178B is similar to constructing a high-quality building. You wouldn't plan "Floor 14" before the foundation. The same applies to planning a DO-178B implementation.

What about a new operating system that's never been used before? You may have heard great things about it and a salesperson said it's certifiable. Nevertheless, it's best to avoid technologies like operating systems that don't have a pedigree, that is, something without a history behind it. You don't want to do a "science project". Pick one of the popular DO-178B certifiable operating systems such as Integrity-178B, then seek other tools that support it. Your operating system is your "brain", so select it based upon what you're asking that brain to perform.

What about digital signal processors? DSP's are used all the time because they're fast. But don't use DSPs just off the foundry line. A lot of COTS are fine but they should have some history behind them. If there is no design information you'll end up with a science project. With DSP's pay particular attention to DO-178B tool support; DSP's are new to the avionics arena, so DO-178B-compliant tool support lags at the time of this writing. We expect that to change over the next few years, but not yet.

While working on a major new airliner, one large avionics company decided to write their own ADA compiler. At the same time they chose to write their own operating system. During development, they kept changing the compiler, while others were writing the application code. But why go through the process of writing your own compiler and operating system? It's a fun exercise, to be sure, but avionics systems should also be developed on time and profitably. Use a compiler that somebody's already developed. The same is true for creating custom RTOS's and tools. Just use tools that have proved their capability in the marketplace. Tools have to be at the highest criticality level as the highest level within your product; the design assurance level. So if the product is at Level A, the tools have to comply with Level A.

Certifying authorities more readily accept arguments for tools and avionics components that have a pedigree, but you should prove it using data provided by the tool company; the cost and time of developing your own data from a third-party tool is significant. Select tools and product components that have a DO-178B pedigree and ensure the artifacts (documents) are available from the vendor.

This is true in DO-254 for synthesis and simulation tools, which are well accepted. They've been used many years and you have evidence from the company of a quality product and CM process. You may have to do a bit of re-qualification to assure that the previous formal tool qualification (as provided by the vendor) is still sufficient, but it's not a huge task to re-qualify a tool. That's why we say go with proven tools that have been through the process.

People sometimes write plans and don't follow them. We see this error all the time. First, the plans are often so high-level they don't define anything---or so low-level they define everything. And some projects don't follow the plans at all. (We call these "practically perfect" plans---practically perfect and perfectly unused. This is a recipe for certification failure). The DER will be looking for compliance in plans that are agreed upon, so it's wasteful to write one and not follow it. People often write perfect plans for beautiful processes they never do. Write what you're going to do, make it clear but succinct, and have it accepted.

**Inadequate level of detail and requirements.** Code should be on the order of 15-25 lines for one requirement. That's a good rule of thumb and, generally speaking, more is better (more requirement detail). In other words, decompose a requirement until there is no room for assumptions by a designer/coder. When the DER or FAA asks for your low-level requirement, follow your traceability matrices and point to a group of low level requirements and say these are my low-level requirements. That's all he's looking for---that you've decomposed sufficiently to where you can write code. Now if the requirements are well defined, and this is true for companies that have a lot of DO-178B and domain knowledge, the low-level requirements have crept up into high-level documentation.

Many companies see significant architectural information or low-level requirements in what you would call a high-level document; let's say it's your SRS, software requirement specification. All the details are contained in it. This is a mistake. What happens next? You need to have low-level requirements.

The engineers next write PDL, program description language, and call that code their low-level requirements. What happens is that it becomes low-level requirements and you have to write a test for every one. You've lost the software life cycle benefit of going through the process of decomposing requirements. It's very inefficient and mistake-prone. So be cautious. Low-level requirements have to be low enough, but don't go too low (like 20 years of philosophical discussion and trade-offs).

**Inadequate and non-automated traceability.** Are you required to have a traceability matrix as an artifact? You are not. DO-178B says you must be able to trace requirements to the code and show evidence of code back to the requirement. How you choose to do this up to you. We don't know how that can be done without a traceability matrix of some kind But you could put documents, say, in an Excel spreadsheet as a management tool. With up to 50,000 lines of code and close to 5,000 requirements you wouldn't try it. It will take a long time to do the conformity audit. Also, an excellent means of proving your attainment of DO-178B transition criteria is the traceability matrix. Use it to show your progression from requirements to code to tests.

Just because you have a traceability matrix, have you done the assurance process? The traceability matrix is a methodology tool in the assurance process to trace requirements to code, and for code to be traced back to requirements. Just having the tool does not mean compliance with the assurance process. This should be automated. It's difficult to have one or two people trying to track all requirements. Get a tool---it can be program-specific but we recommend you use one tool for your entire company and for everyone to learn one process.

**Too many code iterations via inadequate reviews and tool**s. Some people who have not made a checklist write code, go into the lab, then debug it. Then they say, "Oh, that's wrong," so they write more code, go to the lab, etc. That is the way much of the world's commercial software is written. It seems fast because you get quick gratification. Often, managers love it for that same reason; they quickly get to see code executing. However, every study we've read, and certainly all of our own experiences, show that such instant gratification is elusive and ultimately inefficient. Bypassing the requirements and design phase wastes time because you must be sure requirements are right and somebody has reviewed them. World-class iteration of a product, however, is usually two times. Some managers get it right the first time, but in engineering it usually takes at least twice. If it's more than three passes you're doing something wrong---maybe spending too much time in the iteration process. Find a way to eliminate that iteration; it's best to simply follow a good software life cycle development methodology. DO-178B is flexible in that it allows for different methodologies; virtually any modern methodology will suffice. Pick one, describe it in your Software Development Plan, and stay with it.

**Lack of path coverage captured during functional tests**. You know you're going to do structural code coverage path testing and statement coverage for Level C and above. But this shouldn't be done independently of functional requirements-based testing.

You will get 60 to 70% coverage by analyzing your structural coverage during requirements-based (functional) testing. Instead of saying, "Okay, now I go to structural code coverage analysis" you'll be there already. One pet peeve is "formal versus informal". There is no such thing as "informal" for 178B; it is all a formal process. When you're writing requirements, you should be writing the functional test case descriptions at the same time. Likewise, when you have the code and are going test its functionality requirement, perform the structural coverage analysis and gain that structural coverage at the same time.

**Lack of automated testing.** After everything is written and you have the test equipment, the next step is to develop test cases to execute on that test equipment. Doing tests manually is easy; you just go to the equipment, load the software, and push a series of buttons while reading various outputs. But what you want is automation. The number one reason we don't automate most testing is the heavy investment up front. Management isn't receptive because it takes another set of developers for the automation process. So what is done with many programs is to make software changes, then keep rerunning all the tests---which is very human-intensive. Over time, you will almost certainly save money by investing up front in developing an automated test environment. If possible, spend the money up front to get automated because you will save money at the back end of the process. That advice has been circulating in the software world for 40 years and we still don't adhere to it often enough.

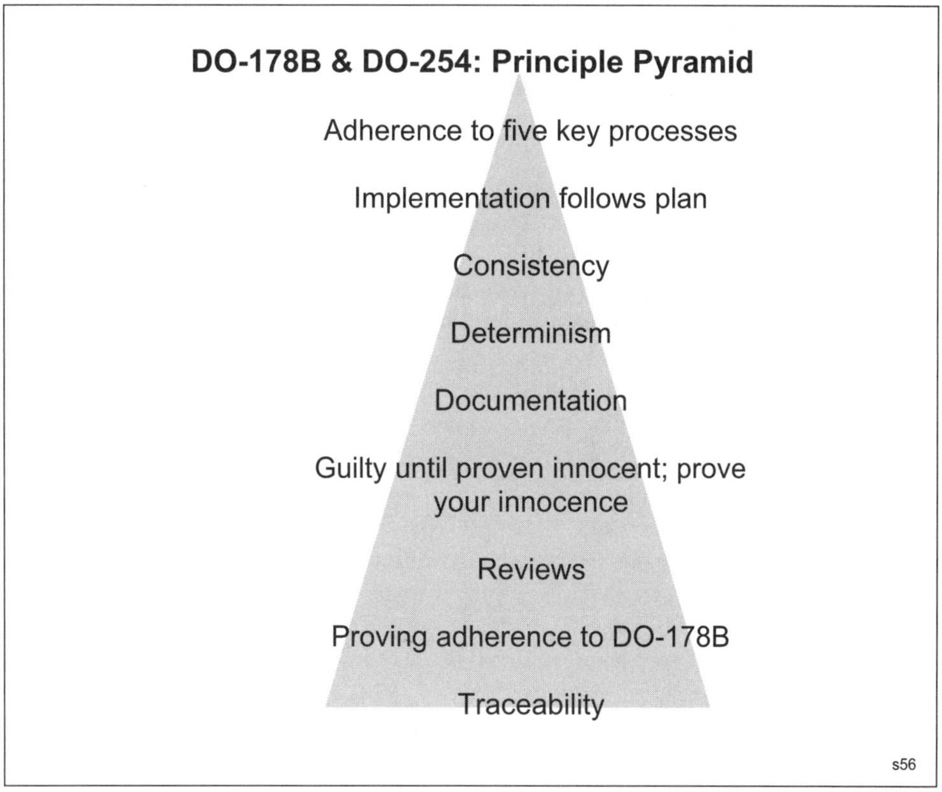


**Determinism**. If you enter and exit a process or test scenario, you always get the same answer. If your process is subjective, then you have some concerns. We like to say the "D" in DO-178B or DO-254 stands for "determinism" but, in reality, you want the same response given the same stimulus and reaction. Your entire software process should be focused on eliminating assumptions; doing so will increase your "determinism".

**Documentation**. When it comes to testing and verification, you must get the same results every time you ask the same questions. Be careful about non-deterministic behavior of a system. For Levels A, B and C, that's a critical item. In Level D, there is quite a bit of leniency, but Levels A, B and C have to be deterministic. Even though we talk about data, documents really build up the data items.

**Guilty until proven innocent**, which is the opposite of the usual court-room law. Here, you must prove you're innocent and did what you were asked.

**Reviews and artifacts generated as a result of reviews**. In visiting two dozen companies producing embedded commercial off the shelf products, we found that most of them wanted to expand market share by entering the DO-178B and avionics world. These widget development companies would say, "We have an excellent process." And the first thing we asked was; "Have you performed the simple task of code review?" This has a lot of value for DO-178B compliance because people often don't see their own errors, even after reviewing the work many times. We always ask that another person examines the data.

Most companies say, "Our code is one hundred percent reviewed before we ship the product." So the next remark is, "Great! You just saved yourself a fair amount of time and effort in the journey toward compliance. Let's look at one review and assess the kind of work done." The response we see is a blank face or "Joe did it six months ago, but he left the company." Thus, there are no artifacts other than "Joe did the code review." We still don't know what was reviewed, the review criteria, how it was done, the findings and how they were applied. It was a halfway discussion. That's why we ensure that the process is followed, with checklists as an artifact.

**Prove your adherence to DO-178B**. This is accomplished by creating tables and matrices that say, "Here are the objectives." There are 66 objectives for Level A and 63 for Level B, which means there's only three differences between A and B. So there's no cause for alarm when the FAA says, "Oh, you're at Level B----we want you to be at Level A." It's only one more objective; not a large amount of extra work. You can, however, present those objectives and describe how they were met and verified. It's not difficult to do.

**Traceability**. The first thing a DER usually does at an audit is look for a traceability document. This is data, Excel sheets or a DOORS database which determine where you are in the life cycle. Did you implement every piece of the system allocated to software? Unless there is traceability and a way of making sure every requirement is oin the code, you can't prove it. Traceability is the scorecard and grade, showing how well the task is accomplished.

It also says, coming from top to bottom, that you've implemented every function. If you go from the bottom up, you show only what is required to be implemented.. There should not be code that you cannot trace. This shows the value of backwards traceability, as well. All blocks of code should have requirements, otherwise you have to ask yourself why it is present and should it be removed.

There are free tools available if you set up your program and processes at the beginning. DOORS is like a Hummer---a huge, powerful machine. It's great for big integrated systems and large volumes of products. But you don't always need to drive to the grocery store in a Hummer. There are tools for smaller projects that are easier to manage. But if you have a large program or multiple development teams and product versions, you probably want to use DOORS.

If programs are small, there are simpler tools to automate the process and eliminate waste. Define size by lines of code: between zero and 5,000 lines is a small-size program; 5,000 to 20,000 is a medium-size avionics software package. Anything above 20,000 lines is considered a large program.

Secondly, look at complexity, not just the system, but the software. How complex is the software? Is it algorithmic intensive? Does it have many conditions? Is there a McCabe complexity number? McCabe is a means to evaluate the complexity of your code, e.g. how many nested "If" statements you have, etc. Many conditions leading to other conditions increase the complexity of software, and complex software is more difficult to review and maintain. So complexity is the second means of measurement.

There is a small, very old book called *Mythical Man-Month* that talks about software. It asks, "What is the ideal size of a software team?" The answer is "one" because the major causes of software errors are assumptions. Multi-development teams are often forced to make assumptions because we software and systems engineers are often not the world's best communicators (just ask our spouses and ex-spouses). Again, documentation and traceability are key items to avoid assumptions.

**The Future**

DO-178C (2009?)

Increased COTS products

Increased scope of DO-178B/DO-254
(ground, naval, commercial industry)

Increased FAA scrutiny

C++, OO, hardware tools

Use of formal modeling/methods

Improved methods for schedule/cost

Clarification from FAA

Industry working groups

S52

## Looking Ahead

We expect to see many COTS products gaining acceptance, particularly in the DO-254 realm, which has been maturing rapidly in the last few years it's been required. The FAA is recognizing more tool sets, cards, CPUs and completely integrated buses. There is greater acceptance of the service history of COTS products and less of a mindset that custom-developed avionics are always better. It's more acceptable to take COTS products and plug them into a system. The more COTS you can justify, the more time you save, but you have to justify the COTS product with a pedigree and application of DO-178B/DO-254 processes.

**Increased scope of 178B.** There will be further expansion in ground, naval, commercial, industrial and other applications. As DO-178B gains wider acceptance, it could become a *de facto* world standard in aerospace and beyond. The Army, Navy and Air Force are either looking at, or already using, it. The authors have taught and overseen DO-178B processes within all branches of the military and each is excited about its growth. Ground-based, naval and Homeland Security systems are also using or considering the near-term use of DO-178B and DO-254.

**Increased FAA scrutiny**. As budgets grow tighter, there is concern about "pencil whipping" (falsifying records). FAA is now is looking more carefully; DERs can no longer just say, "It looks good, I'll sign it." The DER today has to show what was done to demonstrate compliance. A process now used by DERs is SOI (stage of involvement) audits. There are four; planning, design, verification and certification. They are, incidentally, great checklists to be sure a process is working. QA people can also use them for auditing. The DER is now being asked to fill out checklists like these and present them along with their compliance findings. The FAA is also expanding its oversight within areas that were previously considered off-limits: unmanned aerial vehicles and ground-based systems. Several new FAA teams are tackling the military oversight issue, including a new FAA task force working from the FAA's Oklahoma offices that previously focused solely on commercial avionics.

**C++, OO, hardware tools.** 178C is addressing these issues. C++ is increasingly used, although we feel C is adequate. The world is moving toward C++ and today's young programmers will be tomorrow's managers; they'll prefer C++ and tool developers will focus on it. It's no secret that the next version of DO-178, called DO-178C, is in work and will clarify modeling and object oriented technologies, both favored via C++. The same applies to DO-254 and VHDL: there is rapid growth within that area for new tools to ensure compliance. The leading avionics tool vendors are participating in FAA working groups and tailoring their products to ensure compliance. They include Green Hills, Vector Software, Polyspace, HighRely, TNI, MentorGraphics, and Engenuity.

**Improved methods for schedule and cost.** There's more scrutiny where an industry wants to make money The faster we get to market, the better the product schedule. DO-178B does not care about costs; you are allowed to exceed your budget by any known number including "infinity". But the time-to-market is shrinking and the avionics business is very cost-competitive.

**Clarification from the FAA.** There is always new clarification from the FAA and working groups developing methodologies. In DO-248 there are good examples of what is meant by interpretive processes inside of 178B. Regular papers are released and any avionics developer should regularly Google "avionics, DO-178B, and DO-254 white papers" to keep abreast of the myriad changes in this quickly evolving marketplace.

# 9. Safety Assessment

System safety assessments are activities that include functional hazard assessments, failure modes and effects analysis, reliability assessments, preliminary safety analysis and zonal considerations. They create an overall system safety assessment for obtaining criticality level and adjusting architecture based on safety considerations. Programs are fed into the software and hardware life cycle, and the highest level is verified, including system level testing and integrated system level verification and testing. Processes that define development of the system provide plans and life cycle processes for software and hardware. Many of these plans can be reused with minor justification and change.

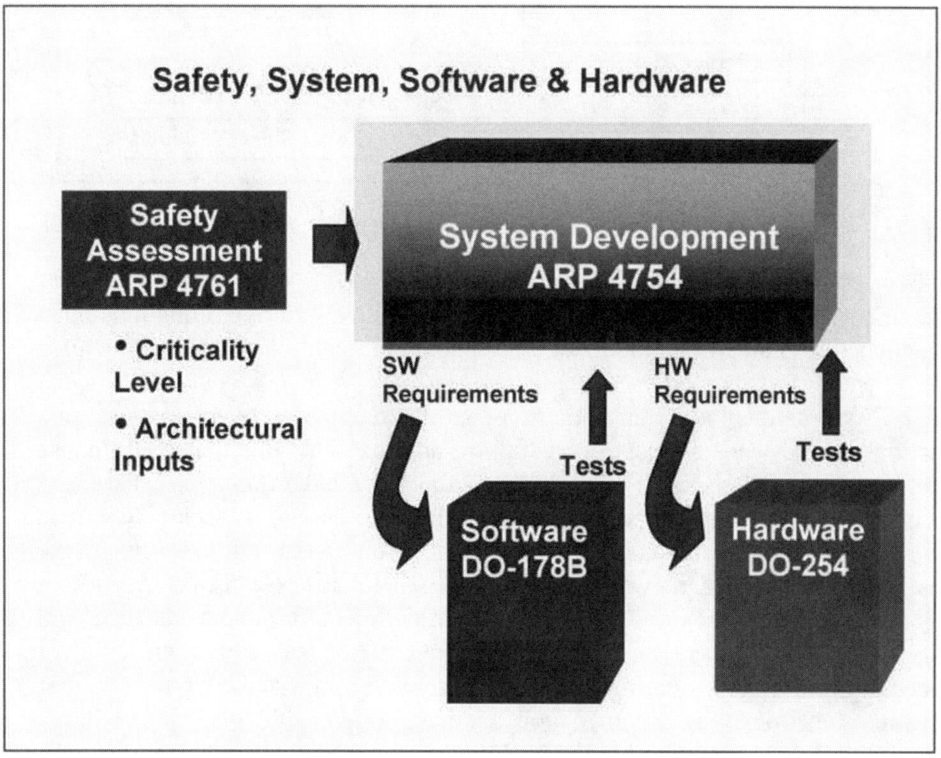

The preceding diagram shows how the safety assessment process is used as a process step prior to defining system requirements. Ideally, the safety assessment should be completed before any system development; however, in real life and real systems this is often an iterative approach. But the end product needs to reflect safety analysis considerations.

The safety assessment process is a series of mini-analyses, each focusing on components and subcomponents within the system. The diagram below depicts a hierarchical abstraction via a set of safety analysis steps applied to a system and components within that system.

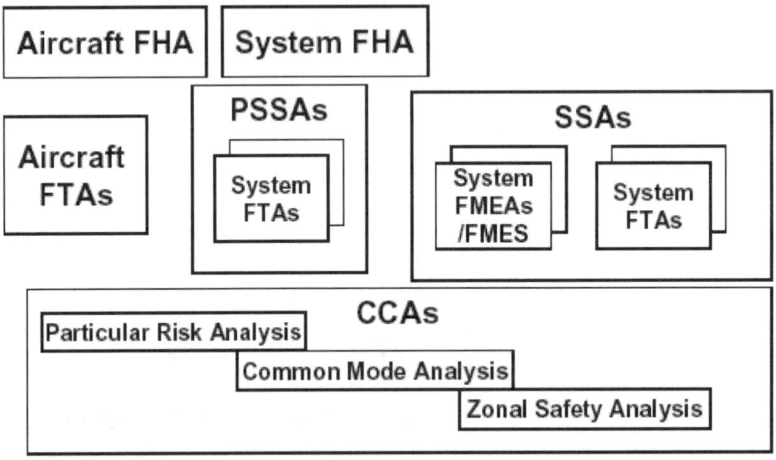


If derived requirements are added, analysis for the impact to the safety assessment process must be performed. Failure points of derived functions must be analyzed, and determination made of impact to criticality level or architecture.

Assume there are many functions within an avionics system, where two functions would lead to catastrophic failure and six functions that could cause hazardous aircraft failures. The remaining functions have minor effect on the safety of the aircraft. They're executing all the time but are different functions and processes on the same CPU. What we like to do is partition these functions. This places a virtual hard wall between the two functions at Level A, six at Level B and the remainder at Level D. Compartmentalize them via time and space partitioning, where no lower criticality function can starve the CPU of processing power for running critical functions or otherwise perturb higher criticality functions. To do that, you are best suited with a CPU that has a memory management unit (MMU). Some processors, such as the PowerPC family, have it---others don't---so choose the correct one.

Also required is a real-time operating system (RTOS) with memory management capability that is able to create walls, or partitions, and detect/mitigate violations. A partitioning scheme for different criticality partitions adds complexity and development cost, but if you want your system to reside within the same CPU, and utilize different criticality levels, you can use the partition system. Why incur the extra work and cost associated with this? There are three primary reasons, all of which have increasing importance in today's systems; reduced weight, reduced power consumption, and reduced size. Today, every airframe manufacturer and retrofitter is increasingly demanding such considerations and, if you aren't doing it, chances are your competitors are or will be.

You can also use a partition system with pin-selectable physical barriers. Software won't even get loaded into flash or RAM without having, say, an external pin or pins connected---you know you're on the ground when those pins are connected. Choose carefully here; partition systems are important.

As we've seen, software criticality divides into five levels, A through E (catastrophic, hazardous, major, minor and no effect). Hardware, on the other hand, is typically categorized into three levels; critical, essential and non-essential. Not every critical system, however, has only Level A software. For example, consider a critical aircraft system with software that does nothing but record hardware faults. It doesn't control anything. Now you may be able to show that the software is running at Level D. Look at the system architecture and be sure this is an iterative process of assuring criticality level and architectural definition to eliminate single points of failure.

## Safety Assessment Concepts

Establish system criticality level

Catastrophic, hazardous, major, minor

Determine design assurance Level (A, B, C, D or E)

Iterate the process to contribute to architectural definition

Use architectural definition to mitigate design assurance level

Safety flows from function(s) provided

Failure, potential failure, of function assessed at all levels of hierarchical abstraction

**Three Key Safety Assessment Processes**

Functional Hazard Assessment (FHA)

Preliminary System Safety Assessment (PSSA)

System Safety Assessment (SSA)

S59

Consider another example regarding architecture and partitioning. We have a partitioned design based on SAE ARP4754. One function is hazardous, the other is major; so one is Level B, the other is Level C. They are independent functions implemented by Function A and Function B, respectively, and partitioned into two different systems. The functional hazard analysis examines Function A alone, then B alone. If both fail, you go to the highest level of criticality. Looking at them individually, a failure of one function can't impact the other because hardware is dissimilar and on separate boards, with separate channels and separate power sources. That represents a well-partitioned design. Remember, Function A was Level B while Function B was Level C.

Assume there are Channel 1 and Channel 2, System A and System B, Function A and B. Both have to operate for the function to work. If one fails, the function fails, so there is some correlation. For this reason, you must be careful to design for the highest level of criticality for communication between them.

## Development/Design Assurance Determination

### Example 1a:

Partitioned Design (Reference SAE ARP4754, section 5.4.1.1, and RTCA DO-178B, section 2.3.1)

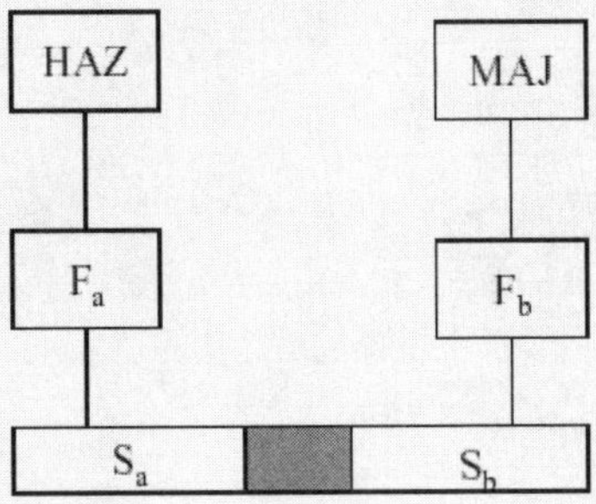


$F_a$ and $F_b$ are independent functions that are implemented by systems $S_a$ and $S_b$, respectively. $S_a$ and $S_b$ are partitioned.

## Development/Design Assurance Determination

### Example 1a: Partitioned Design (Continued)

**SAE ARP4754**

**Level B for the overall system including the partition.**

**Sa is Level B associated to its Hazardous effect**

**Sb is Level C corresponding to the Major effect.**

**RTCA DO 178B**

**If the functions Fa and Fb are implemented in software:**

**Software in Sa is Level B**

**Software in Sb is Level C**
**If the partitioning protection involves software:**
**Partitioning software is Level B**

## Development/Design Assurance Determination

### Example 1b:

**Partitioned Design (Reference SAE ARP4754, section 5.4.1.1 and 5.4.1.4, and RTCA DO-178B, section 2.2.3 and 2.3.1)**

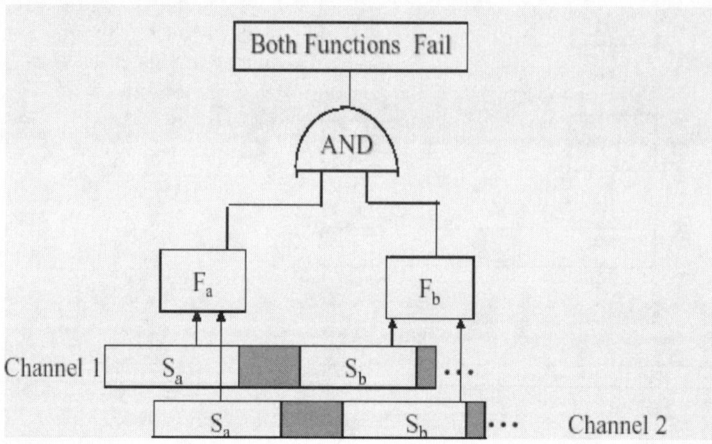


The figure defines levels of design assurance and the probability of undetected failures. For software, that number cannot be calculated; the truth is, the probability of software failure is one, meaning software can fail. It's what happens upon failure that is important to the system safety analysis process and the system itself. Usually, avionics systems perform a power-on self test to quickly check for correctness of the software binary, and this self-test is often performed continuously during operation via a background process. Checksums used to be the dominant technique, but lately a CRC, or cyclic redundancy check, is used. The CRC is a mathematical algorithm applied to the binary, resulting in a unique integer value associated with that exact binary image. The CRC integer is stored within the target binary and the CRC test is subsequently calculated again during operational testing. CRC numbers must match or a corruption has occurred. In this algorithm, if one or more bits change, the whole CRC number changes so it indicates failure.

Each DO-178B criticality level is associated with a probability of failure. As criticality level increases, the probability of failure should correspondingly decrease. The safety assessment process considers all this and more. This book mainly addresses DO-178B and DO-254, so the safety assessment process is not described further, but this process is a vital prelude to avionics software and hardware development and must be fully considered and applied.

# 10. Planning, Development and Correctness

Let's return to the areas of planning, development and correctness, as shown below. Notice how the box "Correctness Process" covers almost the entire area. It's because activities of verification, CM, QA and certification liaison cover the whole life cycle realm, and together comprise "correctness".

**Three Key Processes**

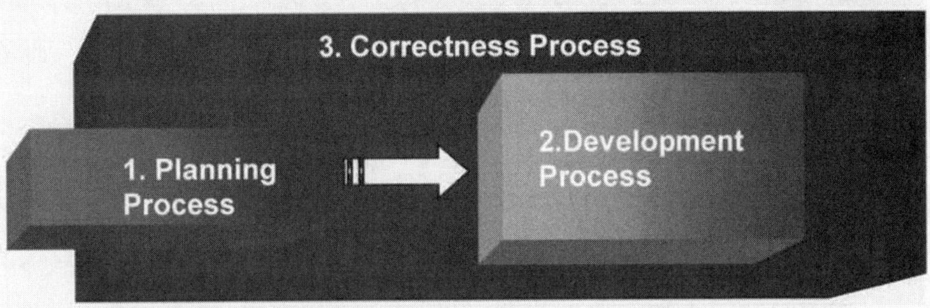


Planning Process first; no development until planning done.

Development Process performed after Planning Process is completed.

Correctness Process performed continuously from project start to finish

The first step is planning; that's why it is farthest to the left. It's the most important foundation activity. It's also the smallest rectangle because it does not require as much time as the other activities. DO-178B requires a review of plans and, if you're at Level A or B, to review them independently. The process contains five plans, plus three standards; PSAC, QA, CM, Development and Verification plans. The contents of the PSAC must cover your overall means of complying with DO-178B objectives, as well as a good understanding of activities for the software life cycle. You provide assurance that you're following procedures and steps in the life cycle that comply with DO-178B. Can you point to other plans or procedures from the PSAC to your company operating procedures? Yes, but remember, once you do they become auditable and can be reviewed. They're subject to critique by both the FAA and your customer.

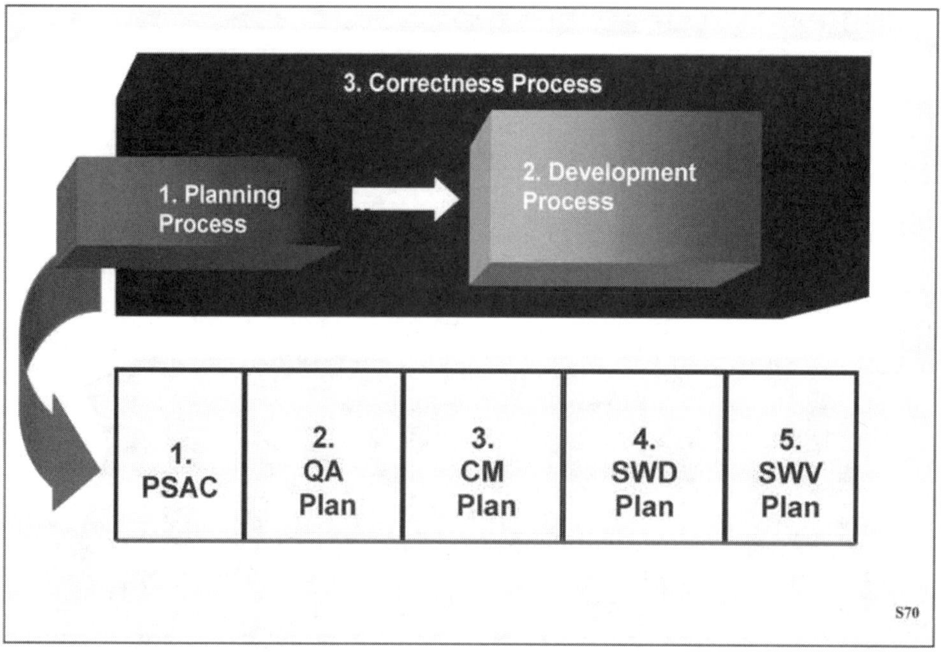


## PSAC Overview

The Plan for Software Aspects of Certification is the primary means used by the certification authority for determining whether an applicant is proposing a software life cycle commensurate with the rigor required for the level of software. This plan should include:

**System overview.** This section provides an overview of the system, including a description of its functions and their allocation to the hardware and software, the architecture, processor(s) used, hardware/software interfaces and safety features.

**Software overview**. This section briefly describes software functions with an emphasis on proposed safety and partitioning concepts, for example; resource sharing, redundancy, multiple-version dissimilar software, fault tolerance and timing and scheduling strategies.

## Plan For Software Aspects of Certification (PSAC)

Keep it under 40 pages

Develop a high level of understanding of software activities

Show compliance to DO-178B objectives

Address specifics with pointers to detail plans

Address special considerations (Tool Qual, COTS, Service history, etc)

Include a top level schedule

S71

**Certification considerations.** This is a summary of the certification basis, including means of compliance relating to software aspects of certification. This section also states the proposed software level(s) and summarizes the justification provided by the system safety assessment process, including potential software contributions to failure conditions.

**Software life cycle.** This defines the software life cycle and includes a summary of each software life cycle and its processes for which detailed information is defined in their respective software plans. The summary explains how objectives of each software life cycle process will be satisfied, and specifies the organizations to be involved, organizational responsibilities, and system life cycle processes and certification liaison process responsibilities.

**Software life cycle data.** This section specifies software life cycle data produced and controlled by software life cycle processes. This section also describes the relationship of data to each other or to other data defining the system, the software life cycle data to be submitted to the certification authority, the form of the data and the means by which software life cycle data will be made available to the certification authority.

## PSAC Contents - Example

| | |
|---|---|
| Table of Contents | Product |
| Scope | Certification Kit |
| Document Identification | Software Overview |
| Document Overview | Certifiable Product Overview |
| Referenced Documents | Product Name Software Components |
| Applicable Documents | User Specified Code |
| Reference | System Libraries |
| Document | Target System |
| Revision | Historical Background |
| Reference Documents | |
| Acronyms and Abbreviations | |
| System Overview | |

**Schedule**. This section describes the means the applicant uses to provide the certification authority with visibility of the activities of the software life cycle processes so reviews can be planned.

**Additional considerations.** This section describes specific features that may affect the certification process, for example; alternative methods of compliance, tool qualification, previously developed software, option-selectable software, user-modifiable software, COTS software, field-loadable software, multiple-version dissimilar software and product service history.

## PSAC Details - Example

Certification Credit Sought for Certifiable Product

Software Development Processes (A-2)

Verification of Outputs of Software Requirements Process (A-3)

Verification of Outputs of Software Design Process (A-4)

Verification of Outputs of Software Coding & Integration Processes (A-5)

Testing Of Outputs of Integration Process (A-6)

Verification of Verification Process Results (A-7)

Certification Liaison Process (A-10)

Software Development and Verification Tool Qualification

Development Tools

Verification Tools

Summary

Software Criticality Level

Assignment of Criticality Level

Additional Considerations

Future Development

Product Name Development

Tool Considerations

Be sure to address special considerations within the PSAC. Even though you may not have previously developed software, put in a paragraph entitled "Not applicable to this program." That way, you've considered it and the FAA doesn't ask; "Did you consider the use of previously developed software?" You reply, "Yes, it's in the PSAC under special considerations. I'm not using previously developed software and am creating all software from scratch, so I put it there."

If you don't put in that paragraph---and it's only a one-liner, you're subject to questioning. So don't leave assumptions. Put in paragraph names and say, "No COTS will be used" or "Use of service history" whether you are or aren't, just state it. And always have a top-level schedule, which will almost always evolve over time.

We've seen PSACs approved that scheduled completion four years ago. The program is still ongoing. You don't go back and update the PSAC because it doesn't make sense, but it might get rejected because you simply didn't put in seven lines of top-level milestones.

The PSAC is an important document because it's the first one the FAA or your customer sees of your capability for doing DO-178B. We've seen companies do a hack job of their PSAC just to get feedback, then follow up and correct it. They don't want to spend the effort on a good job up front. We all know how important first impressions are. If they're bad, the other party will be far more critical of everything. But make a good first impression and you'll stop getting asked for every little detail. FAA officials have told us, "That company---we don't trust them. They delivered their work, just one document half done, and from then on, the project went down hill."

So start with a PSAC that's in good shape. And just don't plunk it down; present it, say, with ten PowerPoint charts describing each paragraph with one slide. They've never read the document and now they're concentrating, which makes it easier for everyone. And make sure your DER or an outside consulting DER reviews and blesses your PSAC before you provide it to your customers or the FAA.

# 11. Quality Assurance Plan

Your quality department's operating procedure is the Software Quality Assurance Plan. For DO-254, it's called the Hardware Process Assurance Plan since hardware is also concerned with processing of development and transition to production of the hardware. Here you address how engineering life cycle processes comply with company plans and standards, dealing with transition criteria and how you intend to perform audits and maintain records. In general, there's one major conformity review at the end of the program but you may want to do another one just before formal verification. QA's involvement throughout the program is very important so determine the audit strategy you're going to adopt because you may be asked to further clarify it. It depends on the FAA individual you're dealing with. Some don't care about which percentage of audits you'll perform while others do. So it's best to include it and avoid the question.

**Quality Assurance Plan Overview**

Provide assurance (write it!) that company plans and standards fully encompass and comply with DO-178B

Provide assurance that all software life cycle processes comply with company plans and standards

Ensure that transition criteria for each software life cycle phase is adhered to

Perform conformity review

Perform audits and retain records/metrics for all engineering life cycle phases

Ensure independence of all QA activities

Make it company-generic: re-use on other projects with minor changes

Keep records of your audits and, in fact, record everything. The trail, or "breadcrumbs," must be in place. The QA plan normally is written by the QA department but if not, they have to approve it because that's their operating plan and procedure. Be sure there are audit artifacts for all categories of requirements, traceability and all design and data pertaining to coding, testing and configuration management.

## Quality Assurance Audits

QA must perform thorough audits

Audits must sample each lifecycle phase and artifact

QA's audits are process oriented, not necessarily technical

Plans, if not written by QA, must be fully reviewed and approved by QA and cover the following items:

All requirements, traceability, design/code/tests, ancillary files and processes

Software test plans, procedures, test cases and results

All third party tool documentation

Tool qualification plans

CM records

Peer reviews

S78

It's important to be able to regenerate the binary, load it into the target and show that it's the same as what was intended. Start from the source files and go through the build instructions. You should be able to load it into the target and see how it functions. That's part of conformity.

## Quality Assurance Conformity Review

QA plans and performs the conformity review to ensure:

All software life cycle data:
- Has complete traceability
- Adheres to company standards and plans
- Is under CM control

All problem reporting is in accordance with CM Plan

Software can be generated (binary), loaded and tested per written instructions

s77

# 12. Configuration Management

The Configuration Management (CM) plan has four major objectives; baseline and traceability, change control, identification of each individual data item and ability to control the version and replicate the product. Within the configuration management plan, specify in different sections the process whereby you are going to achieve these goals.

For the other DO-178B and DO-254 plans (Plan for Software Aspects of Certification, Quality Assurance Plan, Software Development Plan and Software Verification Plan), there are specific persons assigned to performing the life cycle activities defined in those plans. The CM plan, however, is more generic in terms of responsibility; quite literally, every member of the software and hardware engineering teams is directly responsible for CM. Can you have a dedicated person or team solely devoted to CM? Yes, certainly. Do you need to do such, or do most companies implement CM in that manner? No. CM is everyone's responsibility, just like preparing documentation, performing reviews and following the plans; everyone is responsible for CM.

**Configuration Management:**
**Plan Overview**

**Management Objectives:**

Baseline & Traceability

Change Control, Problem Reporting & Review

Configuration Identification

Version Control & Replication

s78

Some companies have two CM systems, one for software source files, and another for documentation. If you can merge the two, it's easier in that you'll have just one system which usually means just one CM tool. But there are some CM tools that do both. It's just that managing software also requires managing documents and data items associated with it.

**Define the software baseline for reviews.** A data item ***must*** be in CM immediately prior to its review. Why? All reviews must designate the version number being reviewed, and most CM tools automatically track and manage version numbers. Only after an artifact is checked into CM can it be reviewed. Can an artifact be placed into CM earlier than immediately prior to its review? Certainly! Remember, DO-178B is flexible. In fact, it's advisable to check an item into CM very early, typically at the end of the first day the originator begins working on it. Why? Prevention of data loss. Ever lost a hard drive or file? Everyone has; CM provides centralized storage and minimizes the potential loss of files, data and artifacts.

It's not permissible to develop software Module A without placing it in CM, then telling a colleague in the hallway; "I'm done with Module A. Go ahead and review it and give me your verification." He can review it and even fill out the checklist, but he doesn't know what version applies until it's in CM. If he knows it's Version 1, he fixes it and generates the next version, e.g. Version 2. Now everyone can keep track and have a baseline prior to reviews. Without a version number associated with the artifact being reviewed, the review is informal and does not meet DO-178B review criteria.

## Configuration Management: Details

Specifies processes pertaining to managing the software and hardware configuration

Defines baseline for reviews and releases

Ensures binaries can be consistently and repeatedly generated

Provides problem reporting, tracking, authorization, incorporation, review and prevention of unwarranted changes

Provides security, backups and recovery of all electronic data (and physical data if no electronic corollary)

Ensures artifacts are controlled, including:

- Application software and documentation
- Plans and standards
- Reviews and audit results
- 3rd party tools
- Software development and test environment artifacts

S79

Be sure the binary can be reproduced, regenerated, and problem reports tracked. You also must have a method of backing up your system and maintaining security. In 1991, a huge earthquake hit the Los Angeles area, causing some suppliers of the Boeing 777 to lose their facilities and data. Fortunately, the suppliers had not only backed up daily and weekly, they backed up offsite, as well---and were able to quickly regenerate the lost data. An important step in company operating procedure, therefore, is regular backup and an ability to reproduce data items. Back up documents, software, test cases, plans, standards and artifacts for reviews.

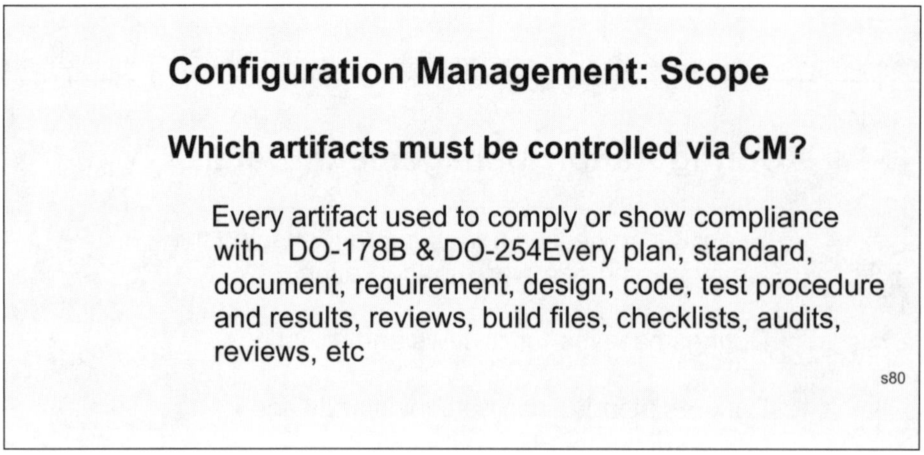


**Third-party tools**. Keep the version number of any tool you use so you can always obtain another copy if necessary and know the environment in which you set up and performed development and testing. Configuration management also asks for control of artifacts. We believe that audits and checklists should be kept in within the CM system as inexpensive insurance and further proof that you have followed your plans requiring such to achieve DO-178B/DO-254 compliance.

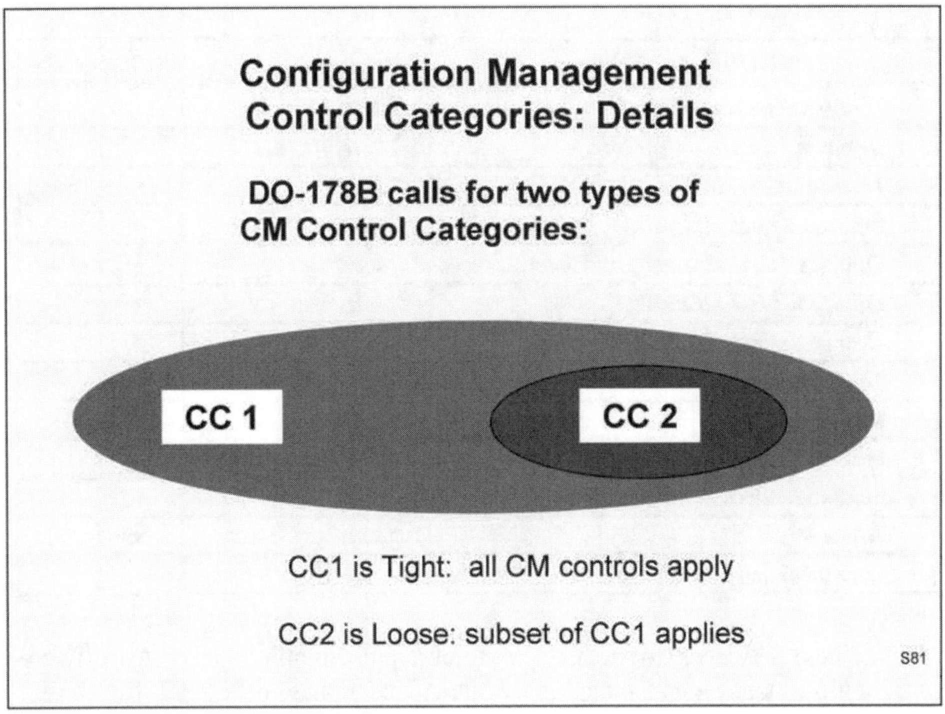


DO-178B identifies two control categories of configuration management. But don't waste time trying to distinguish the level or data item, then finding out if it's got to be Level CC1 or CC2. It's easier to keep everything under tight configuration control so you don't repeat the process and spend more time in determining which level applies than it would take just performing CC1 on everything.

Basically, there are two levels of control; Control Category 1, which is very tight and Control Category 2, which is looser. There are a number of differences, but most of your CM is automated anyway, via a commercial CM tool.

Some items may be checklists. By the time you distinguish between the two control categories, more time is wasted than by keeping everything under Configuration Control 1. The way to identify which data item should be CC1 versus CC2 is to look at the tables in DO-178B that say, for instance, the PSAC needs to be kept under CC1 and these tasks are associated with CC1. This means you need identification, baseline traceability, problem report, release method and data retention, versus CC2 which requires only items with bullets.

SCM PROCESS OBJECTIVES ASSOCIATED WITH CC1 AND CC2 DATA

| SCM Process Objective | Reference | CC1 | CC2 |
|---|---|---|---|
| Configuration Identification | 7.2.1 | • | • |
| Baselines | 7.2.2a, b, c, d, e | • | |
| Traceability | 7.2.2f, g | • | • |
| Problem Reporting | 7.2.3 | • | |
| Change Control - integrity and identification | 7.2.4a, b | • | • |
| Change Control - tracking | 7.2.4c, d, e | • | |
| Change Review | 7.2.5 | • | |
| Configuration Status Accounting | 7.2.6 | • | |
| Retrieval | 7.2.7a | • | • |
| Protection against Unauthorized Changes | 7.2.5b(1) | • | • |
| Media Selection, Refreshing, Duplication | 7.2.7b(2), (3), (4), c | • | |
| Release | 7.2.7d | • | |
| Data Retention | 7.2.7e | • | • |

The PSAC has to be CC1 but let's pick another one. A software configuration management plan for Levels A and B has to be CC1 for C, and CC2 for D. It's easier to keep tight configuration control. To change the configuration management plan, you would write a problem report. The level doesn't matter and you are done. There is one process for all documents and data items for all levels. In practice, people ignore CC2 and apply CC1 to most of their artifacts. CC1 encompasses all activities that CC2 asks for, plus more, so you're covered.

While consulting for some companies, we find that someone might disagree and say, "We want to distinguish levels and make it easier for Level C and D work." They have already established this complex policy which nobody follows and it's a mistake to do it that way. For configuration management, we recommend a commercial tool rather than inventing it in-house---something that's easily understood and followed by engineers.

Problem-reporting tools are available that meet the intent of DO-178B and are easily adapted into processes. If you have one, be sure it meets the intent of DO-178B for tracking changes. There should be manual steps to make sure there are no unwarranted changes. That means that every time an artifact change is reviewed, you compare the changed version and the prior version (prior to the change) and confirm the only changes made to that artifact are those warranted by the problem report; no other "unwarranted" changes were made.

A young engineer with a client company wrote control law software to be flight tested aboard an airplane. While he wrote a data string, he accidentally typed a comma that broke up two data items. During one flight, the airplane began yawing (causing people in the back of the cabin to throw up). It was a mystery why the yaw damper was malfunctioning. The inexperienced engineer couldn't imagine that he did anything wrong until he compared the current software version with one that had been flight-tested earlier. After wasting a week and several hundred thousand dollars, the "improvement" was discovered. It was a simply the comma---but it had been misplaced! Taking steps to avoid those errors can save a large amount of money, and possibly save lives.

## Configuration Management Recommendations

Procure, understand, and utilize a commercial tool to automate all CM processes possible

Procure and utilize a commercial problem reporting tool

Add manual steps to ensure no unwarranted changes are made

Train all personnel in CM; it's everyone's responsibility

Ensure all artifacts are placed under CM (common mistakes: missing 3rd party docs, libraries, makefiles, reviews, audits)

Make the CM Plan company generic: re-use on other projects with minor changes

S84

Train your personnel in CM. Make sure everybody not only understands how to use it but knows it's not a one-person job. Everybody is responsible. Put everything under CM including third-party data items needed for the life of the program and have a generic CM plan for all company programs. Again, why not keep all data items under CM? It's not going to cost anything, it's easy and saves a lot of questions down the road when trying to reproduce records, retest them or re-audit any piece of the artifacts. Remember, your CM system is a focal point for DER and FAA audits; it's a virtual certainty your CM records will bear the brunt of those audits.

## Configuration Management Retention

**All artifacts must be controlled.**

**But which must be retained?**

All? No!

Retain all records showing conformance to DO-178B/DO-254

Retain sufficient records to provide ability to review, rebuild or retest prior versions

s85

# 13. Software Development & Verification Plans

The good thing about plans is that once you develop them, then hire a new engineer, he will know his role in their development. After reading an entire plan, in just couple of hours he'll become familiar with the processes, how-to's and operating procedures for the activities within that engineering phase. In a development plan, we define the relationship between the environment, tools, processes and strategies, as well as the relationship between each phase of the program.

DO-178B calls for two types of software engineering plans: the Software Development Plan (SDP) and the Software Verification Plan (SVP). The SDP covers requirements through integration (including design and code), while the SVP covers reviews and testing.

**Software Development Plan Overview**

Software development environment, tools, processes & strategy

Relationship and transition criteria between phases (definition, design, code, integration, test)

Includes (preferably references) all review criteria (standards and checklists) for requirements, design, code and integration

Covers all software items including BSP, RTOS, libraries, drivers & tools

How is a transition done from requirements writing to design? Or a transition from design to code, and so on? The new engineer doesn't simply sit down and start writing code. Before design, he goes through a requirements review and involves QA. He also needs to reference all review criteria.

## Software Development Plan Recommendations

Specify goals and objectives, rather than detailed "how to"

Reference, not include, standards and checklists (so they can be updated without updating the base plan)

Specify development methodology including requirements decomposition and allocation, design/code technique overview, unit testing (even though not required by FAA), peer review strategy

Make it company-generic: re-used on other projects with minor changes

S87

The key is that review criteria are also defined so it's known what criteria you will be judged against. A set of checklists, therefore, is important. You should either buy them, build them or borrow them. They don't need to be included in the plan, but kept as a separate data item. That way, if the checklist needs upgrading, you don't have to revise the complete planning documentation. But checklists are a good way of measuring how well the document was written. Note that each artifact creator (author) should have the corresponding checklist available at the time of creation of that artifact, and perform a self-review prior to submitting that artifact for peer review. This will improve quality and efficiency of the review process.

**Software Verification Plan Overview**

Objectives and strategies for verification of requirements, design & code

Relationship of functional, robustness and structural coverage and where each is to be performed

References standards and checklists for *all* verification activities

Specifies strategy for hardware versus software-based test

Specify regression testing strategy

Specify satisfaction of "independence" criteria

Describe "transition" criteria for verification

S88

Become familiar with the checklist because it's like knowing the test prior to taking it, and what to study. You write requirements that will pass the test so they should also cover software components (including 3rd party) you bring in, for example; RTOS, BSP, libraries and tools. The development plan encompasses the development environment as well as components brought into actual development. It's much more than the PSAC, which refers to the development plan for many details. Within the development plan, specify objectives and the how-to, up to a level that's detailed enough for someone else to know what to do. There may be references to manuals, for example, for certain aspects of the development environment usage.

You can reference standards and checklists. Specify the methodology; are you using model-based development, structured method or an iterative waterfall methodology? How do you decompose the requirements and allocate them? What are the techniques for coding? What are the techniques for design you're going to perform?

Unit testing may be defined here, even though there is no FAA credit, but you should do unit testing anyway. Developers should always test to make sure that code compiles and links prior to peer review. It's a good step for the developer and peer review strategy.

Peer review shouldn't be done around the water cooler. Make it generic for the company, although pieces may be specific, depending on the tools for the program. There should be a sense of formality with peer reviews; they are important. There does not need to be a formal meeting with everyone present, though value is added by such. But there must be a peer review process that is consistently adhered to per the plan and documented accordingly.

**Verification pla**n. Here you are dealing with reviews, analysis, and testing, and that your verification includes functional, robustness and structural coverage analysis as required by your criticality level. Do you use a simulator, emulator on target testing, system level testing or integration testing? Refer to the standards and checklists for all activities. Look at the checklist to ensure that the plan itself is good, as well as the checklist for test procedures.

Put templates in the plan of test procedure development and test case development. Specify the strategy of whether you are going to do hardware- or software-based tests, whether they require these tests to be done on the target or in a simulator, and show equivalency between the two environments.

How will you perform regression testing? Many companies retest everything whenever they modify the product because the test is automated. It's easier to analyze changes and find the impact. You can plan up front and put it into the plan. When it comes to regression testing or regression analysis, expect a lot of FAA attention. They feel comfortable when they see you've retested every time there is a change or potential impact. Possibly, if the retest takes four weeks every time, schedule and budget may not allow it. But if it's automated you might execute tests Friday night and look at it Monday morning---so automated tests are the way to go.

## Software Verification Plan Recommendations

Specify goals and objectives, rather than detailed "how to"

Reference, not include, standards and checklists

Address partitioning and previously developed software

Describe different forms of verification and the environment for each

Make it company-generic: re-used on other projects with minor changes

S89

**Describe different forms of verification and environments.** You have great latitude in determining which test environment is best applied to which elements being verified. But aspects pertaining to and dependent upon the hardware should be tested in an actual hardware environment; that includes BSP, interrupts, I/O, timing, performance and stress testing. For algorithms, internal processing and derived requirements, a simulation environment may be applicable and more efficient.

Make sure your test plans are complete and the plan is detailed enough for a tester entering the program to understand what he needs to know. The role of verification also needs to be defined as to who is doing what and where.

## Software Verification Plan Completeness

Are plans clear and detailed?

Are roles of verification team members explained?

Are independence aspects addressed?

Are requirements, robustness and traceability described?

Are tool usage, configuration and qualification detailed?

Can a technical person unfamiliar with the item under test execute the test using only the information and directions specified therein?

S90

# 14. System Requirements

System requirements typically are provided by the customer and your marketing department, but their development is quite complex. Requirements can come from multiple sources; preliminary TAC analysis, human factors, operator tasks, operational hazard analysis, system hazard analysis, safety verification and operational analysis. It leads to a set of system requirements and system concept.

Your software must be compliant to fit the system design. Analyze the functions to develop a description, and design the architecture. There are engineers who design system architecture, which usually shows multiple backup processes and components, along with definitions and block diagrams that show all components.

One project we worked on, for example, was a system for motor controllers on a transport aircraft. It's an FPGA for the control of environmental systems. Although a systems design was developed, no software requirements had been developed. The output of the process is obviously system functional requirements data. Typically, we think of this as requirements from the customer, namely, what is the system supposed to do? In some cases, you may have an organization that develops system requirements and design, and validates them . It's helpful to get the actual physical product, or even a prototype, because it's difficult to develop software to a set of requirements that are not validated or proven. Validation of the requirements are done at the system level, as well, not just at the software level.

**Systems Process**

| SSA/FMEA/FHA | Output of Process |
|---|---|
| Determination of Design Assurance process | System functional requirements data |
| Define system functions and requirements | System architecture and design data |
| Define interfaces (ICD) | The actual physical product |
| Analyze the functions | Validation/verification and test results |
| Design and architecture of the system | Problem report generated during process |
| Allocate functions | Systems accomplishment summary |
| Integrate HW, SW and Interfaces | |
| Verify, validate and test | |

S94

Problem reports are generated during the requirements development process, not just in hardware and software development but the system as well. Systems engineers should have practices to trace systems level documentation.

There is also a system accomplishment summary. This data summarizes the accomplishments during the whole system development and validation and testing phases. The FAA has no requirement for a system specification process (like DO-178B for software) so it's referenced in a certification plan for the product, which says what systems will be on the aircraft. And they'll ask for a systems verification that includes ATPs, DO-160 and systems level integration and testing, all of which you need from a software perspective. System high-level tests can be part of your strategy as well as software verification.

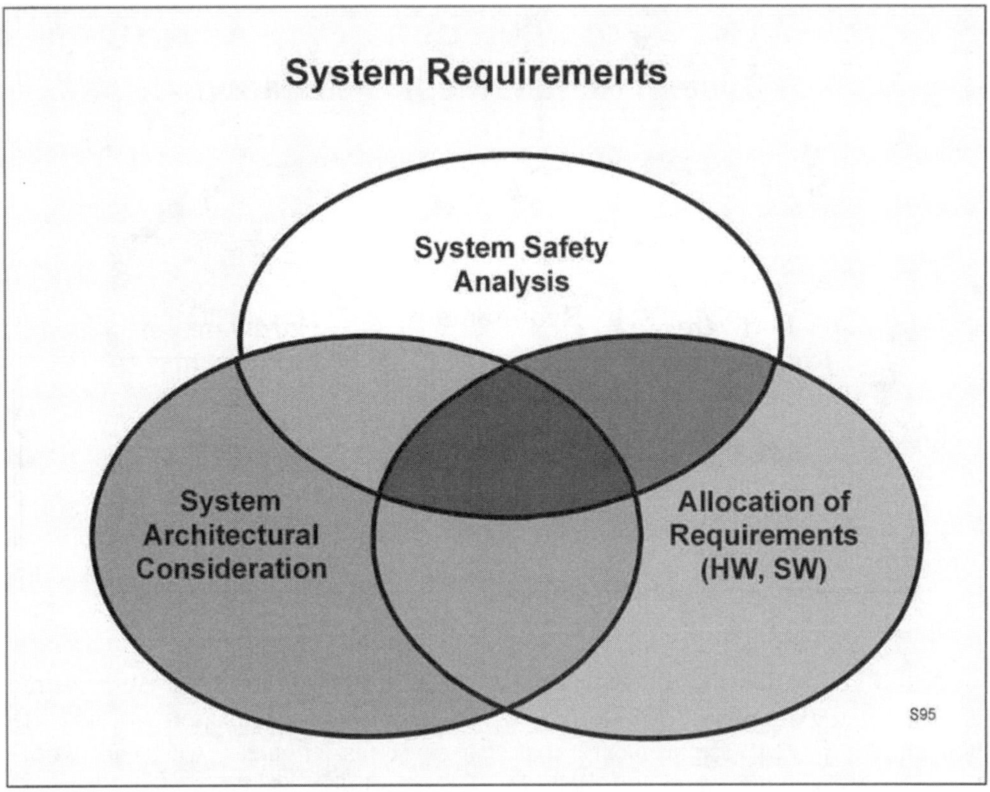


**System requirements detail**. System safety analysis provides us with safety related requirements. Once the two processes are complete, allocations are done for hardware and software. Some requirements overlap, so they need to be traced to either hardware or software, or both as needed. There are multiple ways to do this. Some people just take system requirements, separate and identify them as hardware and software, then come up with the requirements spec. If you're using a tool, that's the best methodology. If you're using DOORS, create another column in your scheme and identify software and hardware, then separate the views. If you're using documents as a process, write separate documents. All those requirements have to trace back to system requirements and design.

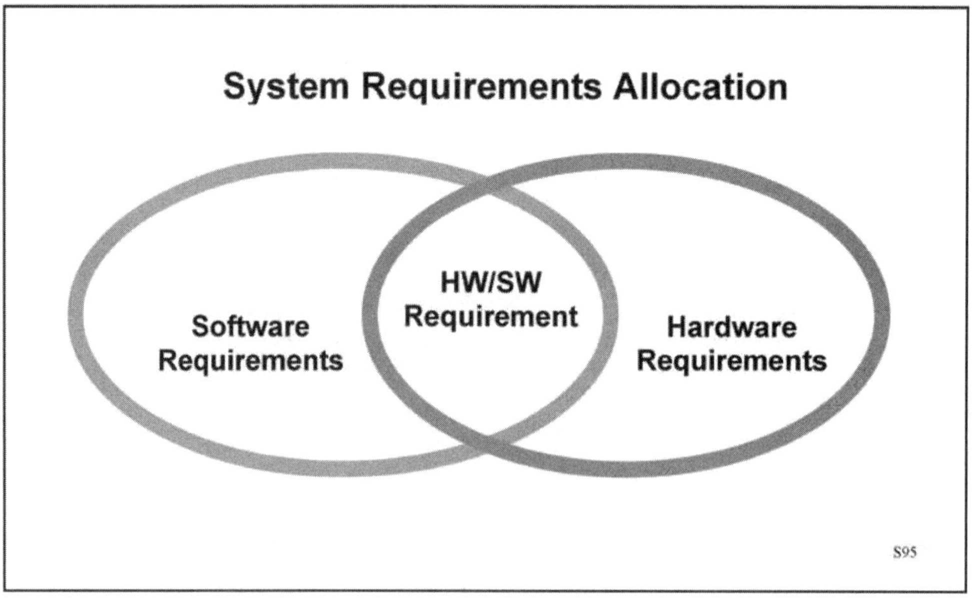


If you're doing an allocation process, that's a transition point. What were the entry and exit criteria? Did you meet those criteria or a transition process? If you find a system requirements writer or systems engineer who can write requirements, pay him more and keep him. He's worth his weight in gold. It is not just scientific methodology---it's an art.

Well-written requirements are crucial. You are unlikely to get what's needed unless you ask for it. If the system isn't specified, it's likely the engineer will specify it his own way. If something should run in a 50-millisecond cycle, state it that way and be specific. Do not use "fast" when you have a specific cycle in mind.

## Requirements Writing

"Well-written requirements are crucial to systems of all kinds. You are unlikely to get what you want unless you ask for it."

*Ian Alexander, Richard Stevens*

"The quality of any product depends on the quality of the raw materials fed into it; poor requirements cannot lead to excellent software."

*Karl E. Wiegers*

Each requirement must be properly elicited, analyzed, documented and verified

Each requirement must be evaluated for desired attributes of its correctness, feasibility, necessity, priority, non-ambiguity and verifiability

s97

Quality depends on raw materials, so poor requirements cannot lead to excellent software. DO-178B starts with requirements and you write software from them. When group organizational theory is done, some systems engineers want to write requirements into the requirements tool. If they can do that, we are well advanced in the development cycle. Systems engineers may want to shorten the cycle of development process by not tracing the requirements or writing problem reports to change the data.. However, this will lead to many problems. System engineers must define and follow a process. Each requirement must be properly elicited, analyzed, documented and verified.

## Detailed Requirements Writing

**Correct** - must accurately describe functionality to be delivered!

**Feasible** - must be possible to implement within known capabilities and limitations of the system and its environment!

**Necessary** - must document real "customer" need or is required for conformance to an external requirement, an external interface, or a standard!

**Prioritized** – is it essential; how essential?

**Unambiguous** – there is only one interpretation of the requirement that can be drawn, either by one or many!

**Verifiable** – ability to devise tests or approaches, such as inspection or demonstration, to determine whether each requirement is properly implemented in the product!

***Requirements from a source must be TRACEABLE!***

S98

Every requirement needs a priority. For good architectural design, requirements for startup receive the first priority. If two reasonable people look at it, there should be no ambiguity, and they arrive at the same answer. If the requirement cannot be verified, it's not properly written. You should be able to write a test for that requirement the minute you see it.

Derived requirements are design decisions that draw on engineering experience but they need justification. These derived functions will be implemented. There will be code in your system that must be traced to a low-level requirement.

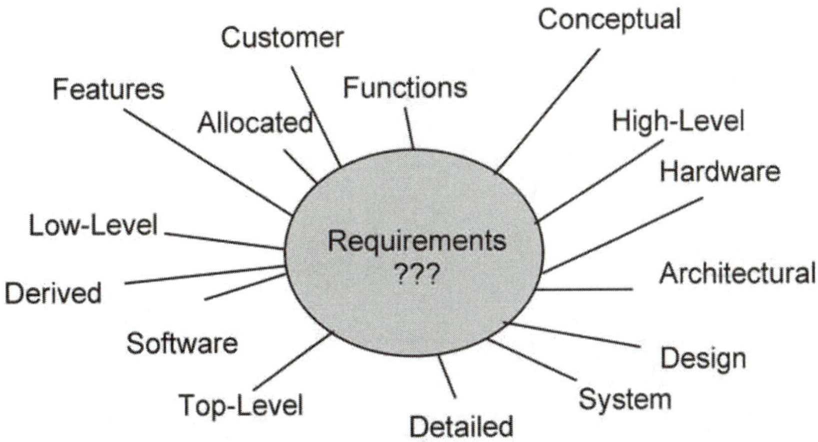


**It's a world of definitions and meanings**
**UNDERSTAND YOUR HIERARCHY**

Assume I have a segment of code and write a design description. Is that a requirement? It depends on who you talk to; some people say it is, while we prefer to say it's not. It's a design description and, left up to the software developer, design must come before code.

Are features a requirement? If they are, it depends on where they're specified. There are allocator, functional, customer, conceptual, high level, hardware, architectural, design and detail requirements. It doesn't matter what you call them, they're all requirements of some sort. If they pertain to functionality or are in the path of generating code, they're a requirement in some form to somebody.

Software has high- and low-level requirements. Hardware has conceptual and detail design requirements. They're called different names but in reality they're the same thing. The point is, it decomposes. DO-178B is unique among software "standards" in that it requires two or more levels of requirements. Most companies just provide two levels: high and low. But you can have intermediate levels if you desire. The key is that you must first consider a higher level of requirements (simply termed "high level" requirements) before you consider lower level requirements. This ensures a step-wise refinement process, as opposed to the big-bang shotgun approach to requirements development. While not foolproof (and writing requirements is indeed an artful science), requiring these two levels of requirements in order, at least ensures that some modicum of refinement will be performed.

# Detailed Requirements Writing

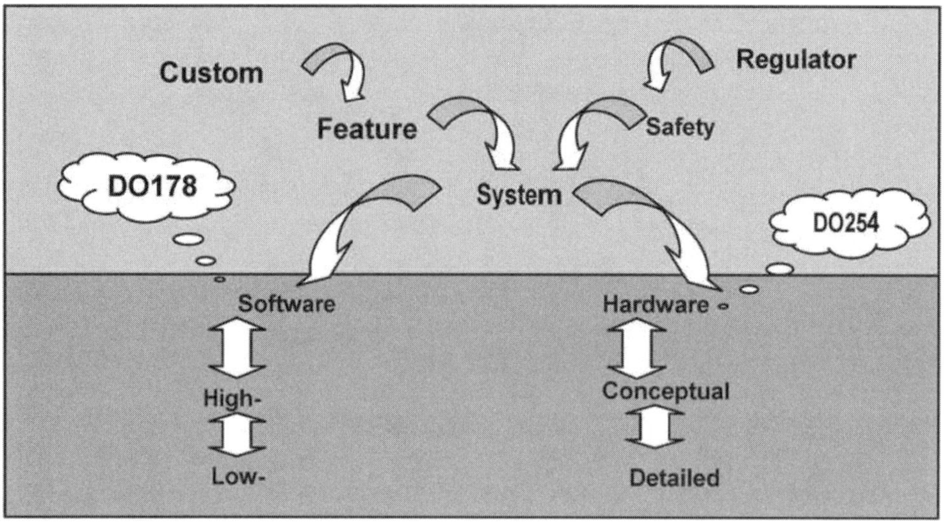


Be careful how far you decompose. Some years ago we worked on the self-defense and display systems for an Air Force bomber. As principal engineers we looked at methodologies and used structured coding and structured specification analysis. To understand why the airplane cost $10 billion to produce, there is a bubble for every line of code. They kept decomposing and decomposing until there literally was a bubble. We went 10 or 12 levels down in decomposition, which is unnecessary, inefficient---and required too much maintenance. Remember our rule of thumb; enough detail so there is at least one requirement for every 15-20 lines of code, on average.

You need to determine the low-level requirement and stop. If you can code from that requirement, stop. Don't argue about whether it's low or high level afterward. No one cares. But remember, assumptions must be eliminated. Decompose until there are no more assumptions needed to be made by a developer. If you can code from features provided by the customer and it's understood and decomposed at that level, you do not have to develop your own high- and low-level requirements. You can stop as long as you can show the flow and trace from those requirements to the code. Both 178B and 254 say that low-level requirements can implement the first code. However, requirements must be identifiable and traceable.

## Detailed Requirements Writing

"The low-level requirements can be used to implement source code." *DO-178B Section 5.2*

"The detailed design data describes the data necessary to implement the hardware item consistently with its requirements." *DO-254 Section 10.3.2.2*

Requirements must be identifiable and traceable
Use of tools to represent graphical or written are a must
Use of tools to identify and trace are a must

Be careful how you identify low-level requirements
vs design descriptions!

**The level of decomposition defines the end-item**

We don't believe that in a modern aviation system traceability can be done without a tool set. We worked with a company who used the word "shall" in a description of a requirement. They wanted to call it a design description but company practices said "shall" is an identification of a requirement. All "shalls" that are design descriptions became requirements and require a test. So they wrote a lot of extra tests. We spent six months figuring out how to get rid of those requirements because they became design decisions. Be careful how you decompose and how deeply you go because it can become very expensive. You must test all requirements but you don't want to write extra, unnecessary requirements-based tests for design decisions.

# 15. Software Design

Software design for 178B is very flexible. It doesn't mandate how the life cycle is done, what documents to combine or methodologies to follow. Its flexibility enables you to define your methodology. Are you going to use cases---will you use UML? Will there be structure charts and data for diagramming? Will you use structured English? DO-178B does say to address four key aspects (shown in the figure below).

**Software Design Overview**

**DO-178B provides for design/documentation flexibility**

**Four key aspects:**

Low-Level Requirements

Interface Definitions

Data Flow

Control Flow

S102

Why does DO-178B provide this flexibility? Remember who wrote DO-178B: a combination of industry and government personnel, all assembled for the singular purpose of finding common ground for avionics safety, and codifying it via a document. That document is DO-178B---and DO-254 for hardware was similarly conceived and defined. Each company at DO-178B meetings would have preferred that their organization's methodology be mandated. However, no company prevailed and many compromises were made. The end result is that DO-178B provides much flexibility in how you accomplish avionics development, particularly in areas of design. After all, there is such variability and evolution in avionics design that this flexibility was a wise choice.

Recall the precaution about stopping before going too far. Somehow, you have to say final requirements decomposition comprises low-level requirements.

Some people say high-level requirements are in the software requirement specs. Low-level requirements are in their architecture or design descriptions. It doesn't matter where you call that break. The truth is the same as real-life; humans are comforted by boundaries, but actual decision boundaries are often gray. If you're using a requirements management tool like DOORS, it's not bounded by your documents. You have a high-level description. There is another description that decomposes and another that decomposes further. The last and middle descriptions can be called low-level requirements or wherever that requirement trail ends.

It doesn't matter what it's called so long as it goes down to the level of where there is sufficient detail to implement the design via code and get it right the first time. Typically, this means a requirement covers 15 to 20 lines of code. But there must be a representation of low-level requirements and interface definitions in the designs. There are external and internal interface definitions within and between hardware and software. For example, if there are multiple tasks, multiple processes, multiple classes and multiple functions, how do they interface? What are the bus interfaces? How do they interface with memory? How do they interface to the operating system? All interfaces, internal and external, need to be fully defined, as well as the data flow and control flow in a design description or design data information.

Do you need a top level design representation and a corresponding low level design description? No, you don't. This is how the military typically documented the design of its avionics systems, e.g. a high level design document and a separate low level design document. But the military is now almost fully behind DO-178B and the greater inherent design documentation flexibility it affords. Likewise, the military is gradually adopting DO-254, albeit at a slower pace, but the authors believe DO-254 will be virtually mandated for all military avionics hardware within the next few years. But the end result on documenting your design will not change. You, like most avionics companies today, will document the design after formalizing the requirements and before implementing code. The document is the design document as described here.

Most of us define data flow in a bubble chart. Or you might have a data dictionary if you learned to document data flow and design in that fashion. But you might use a different methodology, such as the UML process, to define data flow. But data flow is basically information that moves from one piece of software to another in a sequential process. Is it asynchronous, and how is it passed back and forth? What meaning is it trying to convey between software components? What data contents and formats apply to each transaction? These are the questions pertaining to data flow that your design document need to answer.

Control flow and data flow are structured methods corresponding to the analysis and specification of software. Control flow needs to be shown at an architectural level, as represented by a call tree. It goes hierarchically from top to bottom and sequentially from left to right. It's fairly common.

There's a methodology behind structure charts which gives a complete control flow. But somehow you have to specify your methodology and represent the control flow accordingly. For multiple programs, as in a real-time operating system, you need to step that hierarchy up to the control flow. Which programs come in what sequence, and who gets interrupted first? All that architectural control flow needs to be represented, as well.

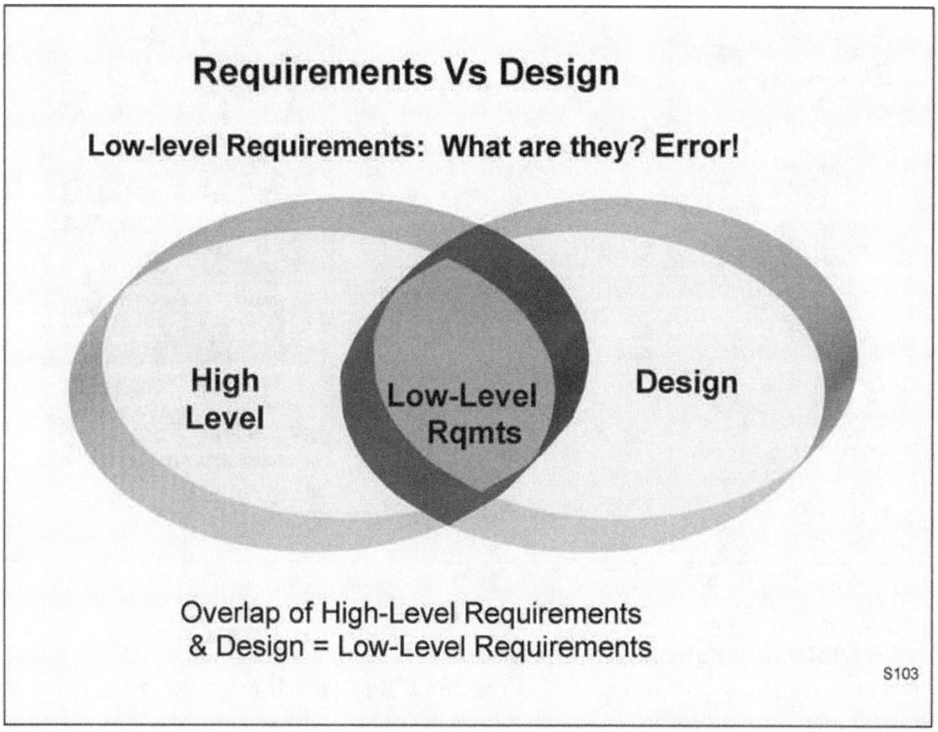


Low-level requirements are somewhere between high-level requirements and design information. You have them when you can code from them, meaning they're decomposed sufficiently. Be careful not to call design descriptions low-level requirements. You'll end up doing a lot of unnecessary testing. DO-178B is less concerned with how you designed your avionics versus "Does your system behave in the manner you specified and does it do so safely?"

## Interface Definitions

**Include all Interface Definitions in Software Design Document**

External, including Hardware
Internal, including all Software:

BSP's

RTOS

Application

Libraries

Drivers

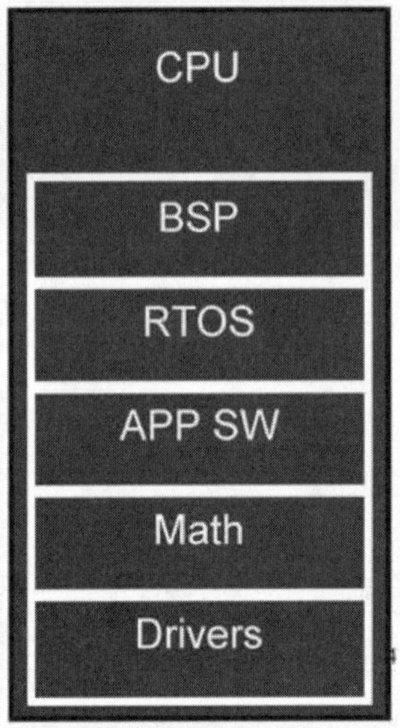


**Interface definitions, external and internal.** The interfaces described here are processed within the CPU. The interfaces which must be specified and documented include ***all*** interfaces between software components, as shown in the above diagram. This includes interfaces between each of these components of software, your application software and all interfaces within your software. Many times we are asked, "But it is my software; why is DO-178B concerned with my internal interfaces?" The answer is simple; yes, it's your internal software but tomorrow it will be reviewed, tested, maintained and changed by someone else. That someone has to fully understand it.

**Control Flow** is a description of how control transitions between modules. Basically, it's your call tree. What is the relationship between modules? Who calls who, when and under what circumstances? Who's hierarchically controlling the lower modules? When are they called or initiated?

When are they sequenced? These questions pertain to control flow and are included in your design document.

## Data & Control Flow

**Control flow is:**

A description of the control relationships between modules

A description of the conditions under which modules are activated

Includes ALL the SW modules

*Thus: A call tree would show all SW modules with all input parameters*

**Data flow is:**

A description of data manipulated within the design

A description of the source and destination for each data

Includes *all* exchanged data between modules

*Thus: A data dictionary would list all data, with format, range, and initialization values*

*S105*

When documenting your design, be sure to include all software modules. The data flow will provide descriptions of data manipulated within the design as it passes through the system software. Include control flow information in software documents describing the flow of external data into the processing stream. Define that hierarchical level; include the next steps to action level. Include the above data flow information in your software design and you'll meet the intent of DO-178B.

## Control Flow

**Include control flow information in software design document**

Flow of external data into processing stream

Flow of internal data within processing stream

Include module activation details for all modules

## Data Flow

**Include Data flow information in software design document**

Flow of external data into processing stream

Flow of internal data within processing stream

Include source and destination information for all data items

Include, or reference, the data dictionary

Include sufficient detail to analyze data coupling characteristics

S106

## Design Coupling and Cohesion (Like Good and Bad Cholesterol)

**Cohesion (Good):** the degree to which functions within a module are related to each other

**Low:** *Bad* — **High:** *Good*

**Coupling (Bad):** the degree to which function's interaction may result in unintended side effects

**Low:** *Good* — **High:** *Bad*

S108

Your software design must consider the two "C's" of software engineering: cohesion and coupling. One way to remember is to relate it to cholesterol in your bloodstream; one type is good, the other is bad. You want to increase one (hopefully, the good one) while decreasing the other. Software design is similar; you want to maximize cohesion while minimizing coupling. We don't want tightly coupled modules, but if you say that in your standard, you have to specify how to define coupling. The same applies to cohesion; you want all functions within a file (module) to perform a closely-related purpose.

## Software Coding

DO-178B provides for implementation (coding) flexibility

Virtually any language, compiler, or style is allowable

Requires adherence to coding standards

S109

Be cautious how you use the language. People have different styles, but coding standards should be concrete in the definition and number of pointers and depth of nested loops. Identify the loop control variable, how it is initialized and how large it can be. Identify how data types are defined and utilized. Identify abstraction and object definition rules.

Consider length of the arrays. How big can they be? What's the standard array size? Is it dynamic? Should it always be fixed? All those questions have to be answered. When you do the coding, however, 178B discusses several issues.

**Software Coding Issues**

Dead Code

Deactivated Code

Code Complexity

Unit Testing

S110

There are several coding issues you must address. DO-178B only lightly touches on them because of their subjectivity. However, we have found that the FAA and successfully-certified projects consistently handled the four primary software coding issues: 1) dead code; 2) deactivated code; 3) code complexity; and 4) unit testing.

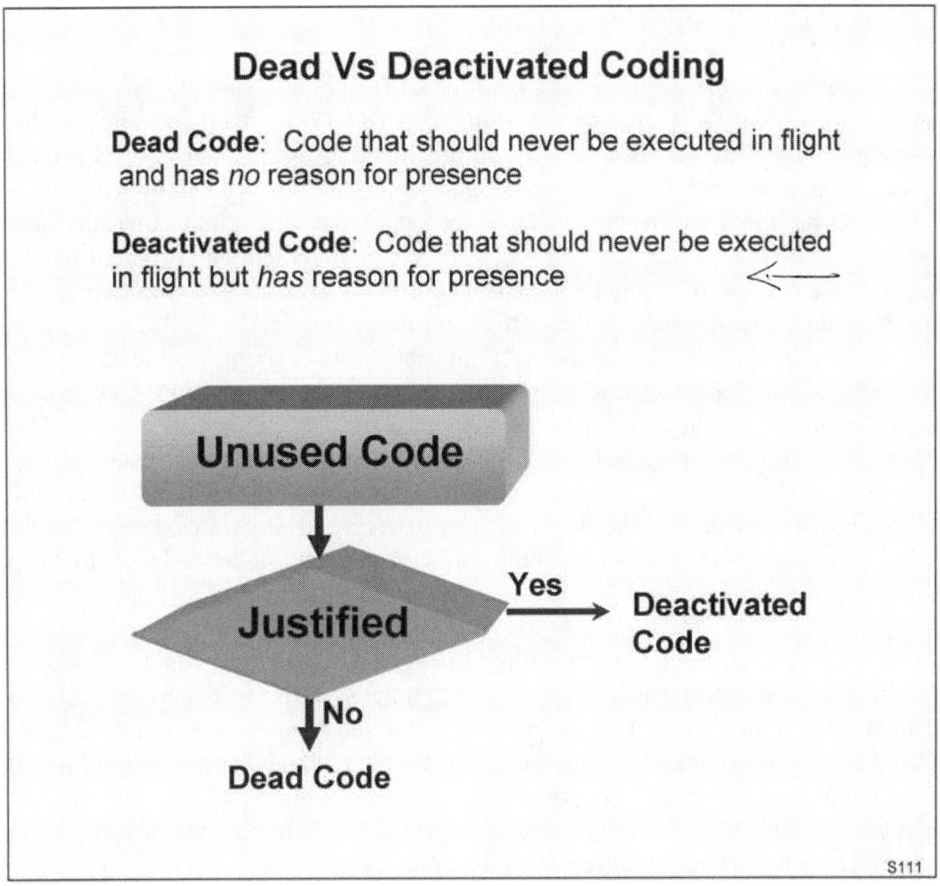


**Dead code, deactivated code, code complexity and unit testing.** It's well known that the simpler the code, the less complex it is and easier to test, review, verify and maintain. Errors are also easier to find. Complexity is usually a measurement of nested decisions, number of repeats and potential segues from the code, like go-to's.

**Unit testing.** However, 178B never directly talks about unit testing, which seems to be a common practice. Unit testing is a leftover from DO-178A, which we don't necessarily think is a bad practice. In fact, unit testing is desirable and should be an important part of your software development process. As purists, we like to be sure every module is working correctly. That, however, may not be the best approach for your software from a 178B compliance perspective. The reason is that unit testing is subjective. Remember, DO-178B is all about objectivity and determinism. Can you define good art or good music? It's relative, isn't it? The same for unit testing; we all know we should do it, it's helpful and important. But as organizations it's difficult to objectively and consistently define, implement and maintain unit tests. Then, should they be rigidly documented? Reviewed? Configured? Audited? Instead, DO-178B focuses on formal verification, post-unit testing.

Here's a question to test your DO-178B knowledge so far. And the answer (of course) is not to be found in DO-178B itself or any related documentation. The question is. "If you are working on a Level A or B project which requires independent verification, and the independent re-uses a developer-authored unit test as the basis for the formal test of that same software, is such allowed?" Ponder that…it would be very convenient and efficient for the tester to re-use the developer's unit test...correct? And it would minimize the need for any assumptions on the part of the tester...correct? Well, DO-178B is not about efficiency but correctness. If the tester re-uses a unit test developed by the code author, then the tester is merely codifying the assumptions (potentially mistaken) made by the developer! So the answer to the question is a resounding "No!" Can you get away with it? Like petty shoplifting in most instances, the answer is probably "Yes." But it's wrong, counter to DO-178B and should not be done. It is not petty---this is avionics software and lives are at stake. DO-178B is about trust, but with ample audit points and independence to keep everyone trustworthy.

**Dead code** is code which should never be executed in flight and has no reason to be present. It must be removed from the system. In other words, if you cannot trace it directly to a requirement or a derived requirement, it's dead code. If you have a bad pointer or a mistake elsewhere in the code, is it possible for that dead code to be activated? The answer is yes, it's possible, and also undesirable. If it doesn't have a reason to be there, it should not be. Simple, but how did it get there? There are two common sources for dead code.

Developers often "experiment" with code during the development process. The end result is some of their work is no longer relevant to the program after a series of evolutions. But dead code is very misleading as it's easy for others, and even the developer, after a period of time, to presume the code was intended to be present. Thus it can be left it and prevent maintenance nightmares, as any large commercial consumer product software company knows.

Also, compilers add code that typically isn't part of your requirements. They'll add small code segments just because the compiler developer needed to make and cover certain assumptions per your source code syntax. But this is not strictly dead code; rather, it's termed deactivated code because your analysis (after performing structural coverage to confirm it can never be reached) will indicate it cannot be removed but also will never be activated. **B**e aware if such code was added, how the compiler works and what it's adding. In Level A, you'll show an object-to-source comparison---and that the compiler added this extra code.

**Deactivated code** should never be executed in flight but there's a need, hence a defined requirement, for it to be there. It flies on the airplane but should not execute. But potentially it can execute. So it's crucial to prove it's deactivated. As an example, there might be a deactivation mechanism such as pin-selectable.

For Level A or B, a single method of deactivation is probably not sufficient. It can't be pin-selectable by itself because it might fail. The code must meet the design assurance process of the criticality for that product, even if it's deactivated. So even deactivated code must be tested.

**Deactivated Code Rationale**

Maintenance software (de-activated in flight)

Different aircraft configurations, via one executable; others deactivated

s112

You're developing a product that will probably work on multiple configurations of the fleet and, depending on the airplane type, certain pieces of software will activate, others will be deactivated. However, if you put B-747 code in a B-737 box, the 747 code has to be tested, as well.

## Deactivated Code Treatment

**What if de-activated code erroneously activated/executed?**

Errant pointer

Apply full DO-178B per criticality level

Requirements, documents, reviews, testing

S113

You need to test for what occurs if deactivated code begins executing. Hopefully, there are byte or built-in tests that detect if you've gone to the wrong place. If a memory management unit is not detecting memory allocations, a watchdog timer should provide a final fault mitigation path. Put in a watchdog timer that is also a path timer. In other words, control should pass certain modules as it executes, and the watchdog timer expects to be affected in a certain way as it goes through the path, which makes it very deterministic.

Apply full DO-178 per criticality level. Deactivated code needs to be at the same level as system criticality at that point. Just because it's not intended to be activated doesn't mean it never will be. You need requirements for deactivated code, documents, design documentation, reviews, testing and verification; all the same information must be created for deactivated code.

## Code Complexity

**Code Complexity**, e.g. McCabe

Relevant to software quality/safety

Like unit testing, unregulated by FAA

**McCabe Complexity Metric:**

An integer value of zero to X, correlating to a source function's "complexity"

s114

Code complexity should be a standard you're utilizing for software quality and safety. Like unit testing, it's not regulated or required by the FAA, but something DERs like to see, especially for embedded, safety-critical code. There should be some standard for complexity of the code. A typical standard is "McCabe," the inventor of a popular code complexity measure. If the reviewer or tester can't read your code, or nested loops are so deep that he's lost track, it's likely that he can't review or test that code properly.

## McCabe Complexity Drivers:

Branches

Loops

Number of entry & exit points

Case/switch statements

Nested levels

Go-to's

**Use max McCabe Value of 8–10; document in coding standard**

S115

Loops inside of loops inside of loops are probably not good practice. As far as the number of entry and exit points, software development standards say there should be one entry and one exit. But you can't always hold to that. That's why we should all know re-entrant code. In deterministic code, you can't be re-entrant.

Case statements can drive the complexity up. In a case statement, what's dead code? It's not even deactivated or dead code; the default. What is it? How do I get rid of the default because you know it's part of my coding standard? My coding standard is going to say I want the default to be present.

In fact, it is defensive or standard coding practice. You may never hit the default in structural coverage analysis because it never hit the default, so you don't have a test for it. Is that dead code to be removed? It isn't; analyze it and see if it meets coding standards. Defensive coding by analysis is approved.

Go-to's are typically not good practice in software development. But if you code yourself into a corner, go-to's can be occasionally useful if done discriminately. Use a maximum McCabe value of eight to ten. It's highly recommended, common and accepted by DERs. It's a quality sign that you're using standards. How will you determine McCabe values for each of your functions? If you had unlimited time on your project and free labor, you could do it manually. Since that is never the case, you could use a software tool such as Polyspace, or Green Hills compilers, both of which automatically provide McCabe values.

## Software Re-use

**Does DO-178B allow for software re-use?**
Indirectly, yes.

**If utilizing re-use, may any steps be skipped?**
Of course not.

**Does software re-use save time or money?**
Usually not
Consider the remodel of an ancient building ...
Only cost-effective if "remodel" is minor, e.g. keeping 90% of baseline intact

**Same for DO-178B Software**

**S116**

Does DO-178B consider and provide for software reuse? Not specifically, but it certainly does. DO-178B was written from the perspective of top-down development of new, custom software. However, Section 12 allows reuse of software from the perspective of previously certified software. Pick up the documentation and code, then go to the CM and QA processes as defined by plans for that reusable piece of software.

When engineers talk about reuse, it's not reuse of the code; it's about copying code. Somebody has written it, and you're doing some re-engineering. 178B doesn't talk about it, but you can define a re-engineering process and be compliant. If you're utilizing reuse, the coding step is obviously shrunk, but don't overlook the 178B steps of planning, requirements, traceability, reviews, testing and development of artifacts. None of those steps may be skipped.

You also go through the process of auditing the reuse package. You have to audit a COTS package ,even if reused, over and over again. Does it save time or money? We, as software development managers, believe it's usually not a good idea. It may take longer to figure out someone else's code, then document it and meet all DO-178B requirements pertaining to it. It's often easier to write from scratch---using requirements and a new design. Managers often don't believe it, but it's an industry fact we've observed via our experience on numerous avionics projects. So the practice of reuse, often thought of as a timesaver, typically is not because in 178B you have to go back and do all requirements, design data, reviews and traceability.

It's only cost effective to remodel when about 90% of the baseline remains intact. If you know what the system does and have all the data, as you would have in the RTOS or library, and the change is minor, then it's an effective method.

# 16. Unit Testing

DO-178B doesn't tell you to do unit testing, or even discuss details about it. For every avionics project, however, we still like to do it. How many errors have been found in unit testing? Many. Think about your own company; you might have gone through many manual steps and contrived many tests. They may be based on software and not necessarily on requirements that led to that software. Typically, we find numerous small errors via unit testing. It can be argued that you know you'll have to do unit testing so the developer is careful up front. He's making sure this unit test will pass. You develop test data and cases for each unit and build the testing relationship environment. The programs are linked with the unit under test so you integrate numerous stubs and drivers to execute this one module.

**Phases of Unit Level Test**

Developing test data and test cases

Building the test simulation environment (commonly stub and driver programs linked with the Unit Under Test)

Executing test cases and determining code and allocated requirements coverage

Evaluation of test results and report generation

S117

Next, execute test cases developed for that unit, perform code coverage analysis----and therein lies the problem. The FAA doesn't like you doing code coverage analysis at the unit level. Why? We want to test functionality of the system and know that coverage is based on real-world scenarios, not a heavily contrived unit test. The components will almost always give 100% coverage because you contrived the test cases by looking at the unit instead of a holistic systems view which could have stimulated that unit in a more realistic fashion. The FAA knows this. Unit testing eventually gives you 100 percent coverage of every statement, decision and MCDC. But it doesn't meaningfully indicate how well the system was tested, or that functional requirements tests are sufficient to perform regression analysis and prove that the system works properly.

## Unit Testing Process Performed by Developers

Extremely labor Intensive

Unit test can consume 50% of development budget

Unit test requires generation of at least 1 test SLOC per deliverable SLOC

Getting ready to test can be 75% of the unit test effort

Not re-usable by testing which is required to develop “independent software tests,” hence not considered unit tests.

S118

In the last phase, there is evaluation of test results and report generation. If there are 50,000 lines of code, with 50 lines for an average function, that equates to many unit tests, perhaps as many as one thousand in this instance. What do you get from them? The FAA doesn’t like code coverage at the unit level and often will not accept it. So what are you really getting out of unit testing? A nice evaluation of test results and a nice peer review of the units. It’s all very beneficial, truly, but at a high cost, especially when considering that the unit test effort is not recognizable or reusable within formal DO-178B testing.

Typically, the unit test process is extremely labor intensive and may consume 50% of the development budget. Think of all those little units; one to go read, another to go write, one to do a square root, another to compare something. Unit tests often require at least one line of code per source line of code, on the average. That’s many extra test cases and data that are not necessarily required. Getting ready to test can be 75% of the unit test effort. You also need test procedure changes and problem reports for every item. Then, consider the fact that source code is often fluid at the time of unit testing, and unit testing is often inconsistent and difficult to maintain in the latter phases of the program. That’s a lot of work, and the FAA and your DER are more than familiar with these facts. DO-178B was written with the intent of eliminating unit testing for a goal of reducing labor while increasing independence.

## Manual Unit Testing: Primary Reasons Why it's Not Recognized in DO-178B

Left to discretion of individual developer; subjective, not objective, as required by DO-178B

Done manually or with home-grown tools

Looked upon as an art form vs. engineering

Very expensive to administer

Difficult to QA review and peer review

Difficult to attach a price to the test effort

Abandoned once integration test is reached

S119

Although unit testing is not required, we still think it has much value to the software development manager. Off the record, we often see more mistakes caught and prevented during the unit test phase than we've seen within formal testing. By having the right test cases, somebody is reviewing those units, looking at the design description and determining if the code was written accurately to standards and requirements. That's part of unit testing. It can be done more as a review process, as opposed to developing test cases and running actual tests. It's pretty much left up to the discretion of each individual.

If unit testing is not done, how is robustness testing accomplished? This can occur at a higher level of requirements and structural coverage analysis. It takes a little more thought, but robustness testing based solely on the software shouldn't be performed in the first place; it should be performed based on the requirements with *augmentation* from the software tester reviewing the source code.

It doesn't make sense to do robustness testing at the unit level or at the code level, except for testing for failures, such as loops, loop counters---and what happens if they go out of range? That kind of robustness testing is still a valuable method, but it's very pricey. Abandoned once the integration test is performed is typically true. After integration testing is achieved at that higher level of testing, hardware/software integration and software/software integration, we've abandoned most of our unit testing philosophy.

## Unit Testing: Manually

Determine closure of unit under test

Determine which dependent units to stub

Build stubs

Build test driver to exercise unit under test

Compile/link test harness-iterative process to correct compile errors

Run test driver main program

Gather test data results and compile into readable format

Build test reports for individual test case and include in SDF

Determine level of source code coverage

Modify harness components for next test case

s120

It's usually a manual process; build stubs, run test driver main program and build test case simulation/stimulation code for each test case. It's not very automated. If you are going to do unit testing, it's highly recommended to automate. It's a good review process. You can use an automation tool like VectorCast from Vector Software; we've used it on most of our projects and it works.

It's interesting that what you call a unit, or small function, you must define as a unit. If it's defined as a large integrated portion of the system that has traceability to requirements, that's different. That's an integration test and 178B wants you to do integration testing from the requirements level, particularly using the target hardware whenever appropriate, which further precludes traditional unit testing.

If you call that the unit, it's highly desirable to do this kind of process. That way, you achieve credit for structural code coverage. Ideally, in structural code coverage, we want to integrate the whole system, build it, then instrument only the pieces you're interested in instrumenting. We like to see the whole system executing. Many times, however, that can't be done.

## Benefits of Automation

Consistent test methodology/documentation

Quick pay back (labor saved and quality improved)

Precludes writing of disposable test software

Satisfies Mil-Std, FAA, FDA and automotive test reporting requirements

Allows testing to begin sooner (stubbing)

Supports target machine testing (VectorCAST)

Removes mundane tasks associated with test

Automates what developers do manually with minimal training

Helps improve SEI maturity rating with automated test process

Makes regression testing at the unit level feasible

S122

It is also good for regression. If it's automated you simply push a button, which precludes writing disposal test software. Also, the test isn't thrown away and is always reusable. To some degree, it can satisfy the FAA, depending on how it's presented.

The key to software testing with DO-178B and other areas is to consider verification and testing in chunks. On many military programs, there is a software test plan; in DO-178 we have a software verification plan that encompasses your review and analysis activities. And it's not just actual physical testing, whether it's on target simulation or any other platform. The thing to remember about verification and testing is that these activities are not sequential. They start in parallel at a lower level of manpower, then grow as more requirements and designs solidify.

If you wait until the end to do verification and testing, there may be a large number of costs and schedule overruns. Many companies tend to staff the verification at the end of the project, and spend a huge effort to catch up. The fact is that costs will double or triple, not for program verification, but the whole project. This is why DO-178B, and DO-254, don't call out a specific "testing" phase within the project life cycle. Instead, it's called "Correctness" and it includes testing, reviews, etc., all ideally performed at the very beginning of the project and then throughout.

You want to contain errors within the phase they're created. CMM/I Level 3 and 4 ask to do that. That's why there is a correlation between CMMI and DO-178B, as well. Start verification planning and execute the plan as early as possible. Write the test cases before you write code, as this helps improve the requirements so the coder will have an even better requirements baseline from which to design and develop. You should also add staff to the verification team early in the program.

Many projects don't follow this waterfall strategy. They already have a working prototype and program in place. Now we're dealing with a planning activity for reverse engineering of artifact development, requirements, design and establishing traceability but, at the same time, starting the verification tasks. So plan the activities accordingly.

Plans should be realistic. They don't have to be waterfall all the time. Many programs may already have a working product. Your plans shouldn't say, "We'll develop requirements and design, then we're going to code." Well, that's not true; you already have a working code. So it's a key area to follow the plan and get verification activities started as early as possible.

# 17. Software Testing

The precursors to DO-178B contained a philosophy that implied quality could be improved via testing. That is truly an admirable philosophy. But ask any experienced avionics engineer with a few gray hairs and you'll be told "testing can assess and indirectly improve quality, but it does not improve quality." And that is the crux of DO-178B; testing (also called "verification") is part of the Correctness process which assesses the "correctness" of each system, and that assessment can improve requirements, design and implementation.

There are four test categories, as shown in the table on the next page. DO-178B prefers you to perform all verification tests at this level, and completely cover the code by running tests at the fully-integrated software level. That's ideal...if we could do it. We've never seen it done for Levels A, B and C. For Level D, there is a chance of performing that. The reason is, for Level D, all you are responsible to show is that you have software requirements that relate to this system requirement and high-level software requirement, and you have performed tests that exercise these functions. You don't even have to do robustness testing, let alone structural coverage analysis at Level D.

There are no tests for low-level requirements within Level D, or to show structural coverage. That's why there's a large delta between Level D and C, primarily due to the simple testing required for Level D compared to higher levels of criticality. Even for Level A, DO-178B prefers all testing be performed at the highest level.

Can you achieve all other aspects of verification if you never "open the box," and just exercise the input to check for corresponding outputs? How can the structure of the code be shown when typically 30 to 40 percent of that structure is, by design, developed to capture errors? Errors cannot be injected readily at a systems level, so there have to be more lower-level tests. Functional tests cover the functions of the box and most functions of software.

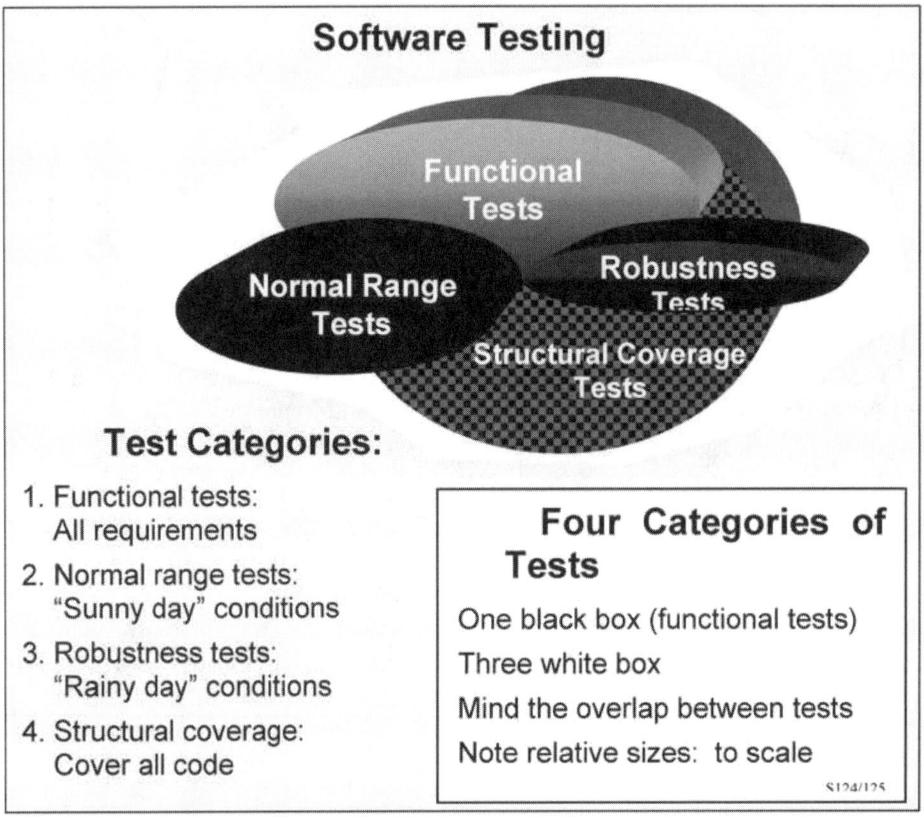


The next level is normal range testing; functions you are designing to. Some actually exercise the software, again at the software level tests. It could be in the simulation or algorithms working the way you want them to. Do they perform as stated for normal operating conditions?

The next level is robustness tests, where scenarios are created that may be in real operation, but occur infrequently. They're "rainy day" conditions where you try to break the software or determine how the software operates under abnormal conditions. This includes error injection which says; "Things that will not occur under normal operating conditions." But we know things don't always go as planned. So we inject errors to go, for instance, for ranges of values, out of range values, boundary values and stress testing. And we're capturing pieces of code as we go along, capturing the structure of functions exercised by doing functional, normal, and robustness testing.

If these tests are done well, you should easily achieve coverage of 70-80 percent of the software, depending on complexity. If good requirements were not written, or low-level functional descriptions and design inadequate, you won't have good tests or good coverage so you'll have to go back and improve the requirements and functional tests by augmenting them. Further, tasks become more expensive because you have to do structural coverage for Levels A, B and C---a tedious repeat of analyzing and verifying code. So DO-178B likes you to push verification tests as high as possible, yet achieve as much coverage as possible because this emulates and verifies real world conditions.

Structural coverage is an analysis of how well you have written requirements and tested them. There's nothing more. It's the exit criteria of verification and testing, the point where you say, "I just achieved 100 percent coverage of software and ran all tests at the system level." That's an ideal scenario.

Now you can stop. You don't have to perform additional testing of yet more conditions because there are actually an infinite number of conditions and you could, ostensibly, test forever.

An earlier version of the document, DO-178A, didn't emphasize this level of testing. It said, "Do functional tests and normal and robustness testing. And then perform structural coverage analysis via additional tests to test previously untested structures." This meant you had two sets of tests. Unit tests can cover all structural coverage because you're exercising execution of the software. Many companies were tempted, and actually deployed, separate structural coverage tests via outsourcing to less expensive structural coverage testing "houses." But this form of testing rarely finds problems, hence the code isn't really "covered" as desired within DO-178B. Separate structural coverage testing of the code merely exercises the code as written, with little regard for why it is there for the real-world.

DO-178B's structural coverage is an analysis, not a test. Once you've analyzed how well other tests were created, you augmented and filled in pieces that weren't covered. And the key goal is zero structural coverage tests because you analyzed all your test coverage reports---and you're done. Don't be concerned when you hear "structural coverage," because it's to your advantage to know when to stop testing. You've achieved the exit criteria.

A question we hear is, "I've written all functional tests I can possibly conceive. And I've generated scenarios based on data to create robustness testing, which means I'm trying to exercise code outside the boundaries of normal operation to see if the software or system can handle it. And I still haven't achieved 100 percent. I'm at a loss. What's missing?"

First, go back to your requirements. The reason you're not achieving 100 percent or close to it, is because you didn't have good requirements for writing test cases. Write more requirements that map into the functions and ensure all your low level requirements are present, traced and verified. The code is there and there has to be a requirement pertaining to it. So line up your defenses to add more requirement-based tests based on the functions.

How much robustness is enough? There is no good measure here. Unfortunately, DO-178B is somewhat vague, as it has to be since robustness tests are largely determined by both the requirements and the implementation (code). If DO-178B were to be an all encompassing document, it would have to address all possible types of systems, designs and implementation, and it would be tens of thousands of pages long. The end result would be that no one would read, let alone follow, it. So DO-178B had to be somewhat vague on these subjective issues of which robustness testing is one.

For proper robustness testing begin by identifying tests you are considering which "break" the software; these are robustness tests. That way you can prove to your auditor at the FAA that you actually generated robustness tests and have a record of them.

Secondly, it's not required to have 10 percent robustness tests or 30 percent. You just need to have addressed them as appropriate for your requirements and code. So, ensure that all of your requirements which have potential boundary value, error values, min/max values, etc, are addressed. Then do the same by reviewing the source code and further identifying such tests. These are your robustness tests. There's no rule of thumb that says, "In this system you should have 30 robustness tests, 50 functional tests and 25 normal operating conditions." Such delineation is impossible, given the many different systems and software. In fact, within your own system it is virtually assured that some requirements and modules will have many robustness tests, while others have few. And these are all within your single system. Black box and white box testing have overlapping normal and functional tests and robustness, and functional tests that tend to overlap. There will be a certain amount of repeats and overlap within these areas. Overlap is good; your DER will review such overlap to ensure thoroughness. Without overlap, it's a certainty that you are missing some tests.

Some portions of the software won't get exercised with those tests. They can be analyzed and then you can add additional structural coverage tests. They're like a unit test, just trying to exercise the code, but they're developed independently.

What's the difference between black and white box testing? In simple terms, black box testing is when you can't see inside the box. Inputs and outputs are known but not the activity within

Because requirements specify what the box should perform, you know what the output is. White box testing is when you review the implementation and select inputs and feed back into those inputs. Or within the actual operation of the box, I can put jumpers or input simulations of data I know are erroneous.

I can also enter a function that's not readily accessible by a combination of inputs, but I can go into a single set of functions or module, force an input that would have been created and exercise the software function in that fashion. We don't like this approach, and like to minimize it, but sometimes there is simply no other alternative.

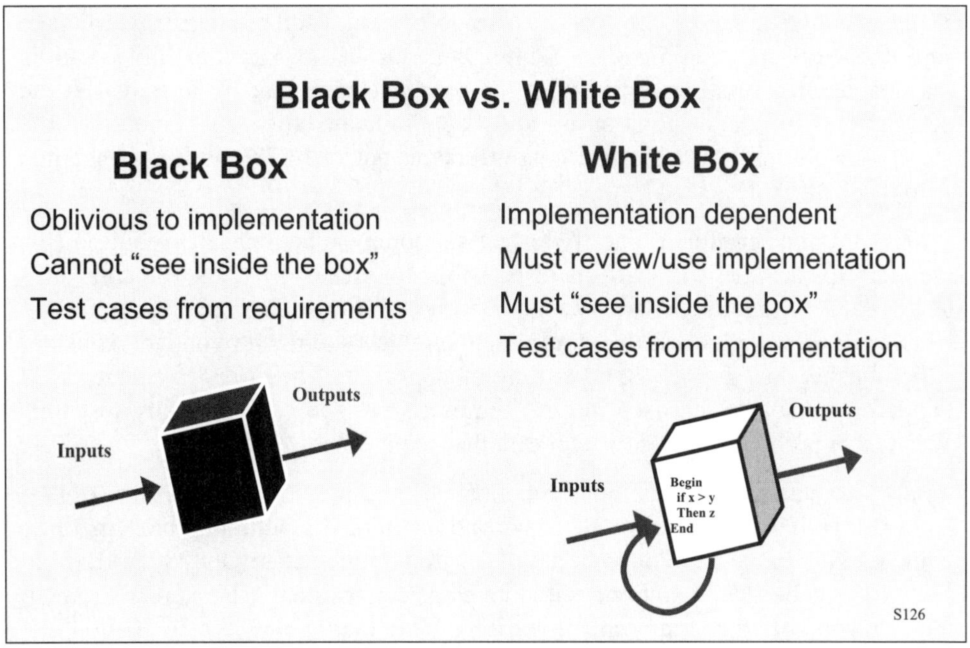


We write software from a high-level spec, requirements of software, low-level and design data. That's where white box testing covers low-level functions, which weren't obvious when they were developed from a high-level system spec. They were needed to write the code and test it, as well. It's a one-to-one correlation in many ways.

Some requirements can be tested on the hardware/software integration test platform, which means, on the box, they represent your target platform. It doesn't have to be the actual box; it could be a motherboard or emulator, but it needs deliverable hardware interaction.

The tests depend on functioning hardware. You can't simulate it via a non-target environment (host, simulator, etc.) and say, "It's going to react in the same fashion." The timing may be different. So it's important that these are hardware/software integration tests. Software integration tests are functions of software that are dependent on each other, but not necessarily on the target platform. So it's a software/software integration.

Low-level requirement tests are normally functions unique to the design or architectural consideration and are categories of verification tests we generate. We look at how much of these requirements were covered. We're not measuring code yet, just looking at it, asking, "How well am I covering requirements I've written? There are, for example, 425 requirements and I've written 1,500 test conditions in multiple categories. I've covered every one of those 425." Or I'm not covering 25 of the 425 based on 1,500 tests. I will add another 60-80 tests to cover those 25 conditions and directly trace to them." This is done because requirements may not be complete. I also go back and add requirements to make sure the tests are complete. Remember; DO-178B has flexibility and is reasonable. It is less concerned with the path you took than your arrival at the destination. And the destination is safely deterministic software. Hopefully, it is also re-usable, maintainable and cost-effective, but DO-178B is less concerned with those attributes.

From a schedule point of view, it's recommended that after requirements are started, the verification team responsible for traceability, then testing, also gets started. It doesn't need to be a full force of 30 test engineers. There might be two people who look at the requirements, review them with you, and consider which bucket they go into for a verification approach. They need to consider, "Is this even testable? Stop writing the requirements...we can't possibly test this. We'll have problems unless we work on that requirement."

So start writing test conditions as early as possible. There may not be code at this time and probably there will not even be a functioning box (that box is most likely being developed in parallel with compliance to DO-254!). But the analysis can be done up front with the verification team. Be sure to remain independent of the requirement writers. At that point, you should feel comfortable that you have adequate requirements and tests.

Also ask, "How much of the software did I cover?" A tool should be used if there are more than 3000 lines of code. You don't want to do it manually. Based on functional and normal robustness tests, you've determined that you've covered 92 percent of the code. Or, you've covered 78 percent. Remember, there is no magic number of goodness or correctness. Simply put, you should have covered "most" of your code at this stage.

In the DO- 254 world, tools are not yet as mature as we'd like. DO-178B tools appeared in the early 1990s and are not perfect today but much improved. (Remember, there is no such thing as a perfect tool.) A single tool may not have all the features you need, so the job might have to be done by several tools. The brochures of tool vendors sometimes say more than they deliver, so examine them carefully to see what fits your environment.

You may become aware that some portion of code never had requirements written for it. When requirements are added, tests become incomplete and need to be augmented. Measure again and you may reach a higher percentage of coverage. This is one of the most expensive aspects of the project. It runs anywhere from 35 to 50 percent of an overall budget. Try to keep it closer to 35 percent, if possible, 40 percent, at most. You're not trying to improve quality by testing, but prove functionality, and that you performed everything asked of you. Quality comes from other activities in the life cycle, such as correctness, reviews and analysis. Structural coverage is probably the most inefficient and expensive ways to find errors, though it's useful; it's just an analysis to tell that you're done.

Look at the number of errors based on structural coverage testing itself. You'll find you spent a lot of money for a minimal number of errors. But those errors were important to uncover.

**DO-178B Software Testing: Verifies and Indirectly Improves Quality**

**Testing is an expensive phase in which to find problems**

Inefficient method of improving quality

"Correctness," not just "testing"

Structural coverage is very inefficient...but necessary

Highest quality when against integrated software

Unit testing is not asynchronous or real-world; many assumptions

S128

A question often arises, "How do I show traceability of structural coverage tests?" The structural coverage test has to point from tests you develop to code segments you covered. You do not have to trace back to the requirements because your test already traced back to them. Trace existing tests based on requirements for the code functions. You probably will not want to go to the line of code with each test. So the higher up the testing in the requirements chain, the more value it has, because it's more realistic. If you test everything at the code level, and just write test cases based on the code, you won't catch as many errors as on an integrated testing level.

**Functional Testing = Requirements**

Which Requirements?

Software requirements

High-Level requirements

Low-Level requirements

Method of Test = "Test," not demonstration, inspection or analysis

S129

DO-178B asks us to ensure coverage of functions we're going to code and perform. They include high- and low-level requirements and analysis in some areas. But as much as you can, develop tests for them. Testing is the only means to be truly deterministic. In other standards, such as former military software testing standards, emphasis is also placed on Inspection, Analysis and

Demonstration as acceptable testing means. But ask yourself; can you objectively and deterministically develop consistent tests when the sole method is inspection, demonstration and analysis? You can argue and ponder but the answer doesn't change; No. Inspection, Demonstration and Analysis are difficult to perform consistently. So DO-178B emphasizes Test as the proper method to conduct verification.

Tests could be on different platforms, but you physically prove it that way. Analysis and demonstrations are a one-time shot, and they're manual. They need to be repeated every time there is a modification.

The more black box testing you can perform, the better. The more coverage based on black box testing, the better your credibility with the FAA and your DER, and the quality of testing. The key is, what is actually used on the aircraft?

> ## Equivalence Classes
>
> ### Example 1: valid voltage range of 24.0–28.0 vdc
>
> **Questions:**
>
> Must you test all possible values between 24.0 and 28.0?
>
> What are the boundaries?
>
> How many in-range voltage values must you test?
>
> ### Example 2: the Sine function
>
> **Questions:**
>
> Where does the sine function come from?
>
> Do you have the source code for it?
>
> Do you trust it?
>
> Do you think the FAA trusts it without verification?
>
> S130/131

Equivalent classes minimize the amount of testing. Say we have a function that acts the same over a valid range of 24–28 volts. Eliminate many of the test cases, reducing costs substantially because you don't have to test at 25, 26, 27 and 27.6 volts. Pick a value between 24.1 and 27.9 and show its equivalence in testing and functionality. You're not going test every point between them. However, you need to do the boundaries at 24 and 28. And we'll see robustness tests just below and above the valid boundaries. That's robustness testing.

Equivalence classes eliminate repetition of the same test with the same results every time. When there is a range, you don't need to test every value.

Consider, next, the sine function and if there is no source code for it. What if it's just a function brought in by the compilers via a binary library function and I just utilize it? Can I trust its value? In this case, the sine function is easily tested using equivalence values, where you show that selected inputs in range of the sine function provide the expected values. Test cases can be created at the boundary points, and the sine function is covered throughout its 360 degree range. However, you don't need to test every degree. Test several values including the asymptotes and negative values, and show by equivalency the whole function is verified. The FAA needs to trust that it works and it's verified in this manner. Many libraries now come with qualified products.

**Equivalence Classes**

Y = sin(X)

Since you do not have code, you must rewrite it from scratch, correct?

No! Equivalence Class it

S132

The original "qualified" compiler was ADA. The terms "qualified" or "certified" compilers, that appeared in the late 1980s, are used by DOD, which mandated ADA. Scrutinized for their compilers, ADA companies generated a benchmark set of code and knew the exact output after compilation. There were tests to run against the output, which looked at it very cleanly. If your ADA compiler needs to be qualified, you ought to run the same sort of benchmark code. If you get the same results, you show you're qualified.

Obviously, run-time libraries need to be verified and qualified in the same manner. However, compiler qualification/certification is not 100% accurate. No compiler benchmark can cover the infinite range of potential source code constructs. DO-178B recognizes this and simply ignores compiler qualification for being the inexact, artful science it is. You get no credit in DO-178B for qualified compilers. You still have to completely test the output of the compiler, e.g. the binary, via the full complement of DO-178B required tests.

There is also the certification package for libraries. In most cases you will not have the source code, but only the binary within the compiler-supplied library. So you normally have to perform source code reviews, functional testing, structural coverage analysis and all the other verification activities called for in Section 6 of DO-178B. However, the FAA knows that in many cases pre-existing COTS software is of sufficiently high quality and, in fact, may be safer than continually rewriting certain library functions from scratch. But how do you achieve DO-178B compliance when you don't even have the code to review or cover? Use equivalent classes to verify your function, along with alternative means of compliance and perform your tests that way on the applicable elements. You have shown it is working as intended. You still perform testing on the code, but use alternative means of compliance to avoid all the activities normally required by DO-178B since you have little alternative other than writing entirely new libraries from scratch. Keep in mind that DO-178B was written with significant input from aerospace companies and those companies do not advocate or desire wasting money re-writing when it doesn't make sense.

## Equivalence Classes

Consider: Y = sin(X)

Comes from compiler- provided math library

You do not have source code

No problem: exhaustive testing via equivalence classes with alternative means of compliance:

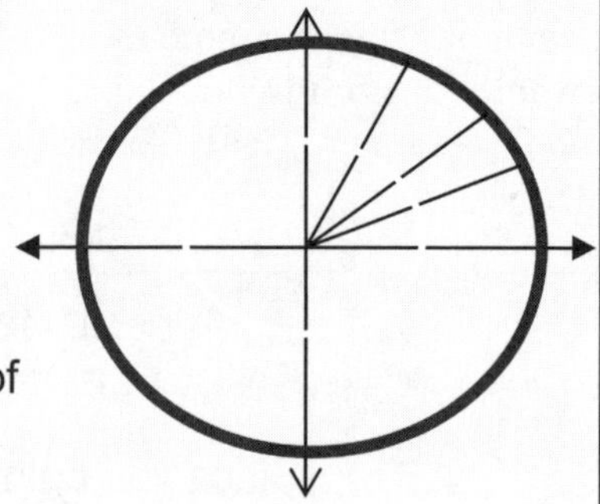

- Test boundaries at asymptotes
- Test several representative values
- Claim, and receive, *"exhaustive testing validity via equivalence classes"*

S133

## Normal Range Testing

- White-box tests, e.g., open the code
- Normal operating conditions
- Normal range values
- Equivalence classes
- Example: the Sine function

*Note: the vast majority of normal range test cases should have been covered via functional testing*

*Why?* Detailed low-level requirements

S134

Normal range tests consider normal operational conditions of the system. Use equivalence classes to reduce the number of test cases and still cover the full functionality. A large amount of software can be covered just by running normal test cases. By powering up the box, testing the power-up sequence and getting the system to react by saying, "I'm alive and connected," you may hit up to 20-30 percent of your code, as long as you verify all pieces as they are activated.

Normal range test cases should have been covered via functional testing because functional tests cover high- and low-level requirements. You may add a little more to cover pieces not covered via functional tests to cover the additional normal fringe.

## Robustness Testing

White-box tests, e.g., open the code

Abnormal operating conditions

Boundary values

*Which boundaries? At and beyond bounds!*

*Bit-level accuracy on boundaries*

Example: the 24 VDC–28 VDC requirement

S135

Robustness tests, on the other hand, are where you create normal scenarios the system may see during adverse conditions. For example, wrong inputs, corrupted data and boundary values you thought would never be reached. But what if it happens?

Some testing becomes particularly important if there is, say, partitioning. We start inputting, for example; we cannot put enough load on the electrical system that disrupts the power, the lighting system or video/audio presentation. But we have to try all scenarios that realistically could happen and see if there are disruptions. Partitioning requires an ample amount of robustness testing.

Again, look at boundary values of each data item and verify at and beyond those boundaries. Why create such things as a data dictionary and fault trees? Because then we can more readily examine the normal data operating range and descriptions. The normal range, then, needs to be verified outside of its normal operating condition to verify the impact upon the system.

Back in 1999 a probe sent to Mars was lost because navigational commands were scrambled---with a loss of 125 million dollars. The reason was simply a unit issue; a developer used English units, while others used metric, and these elements needed to communicate with each other. The error should have been caught during code reviews, data reviews and architectural reviews. If there were a data dictionary, it would be apparent that a particular variable function had to follow the correct units.

During normal and robustness testing, especially robustness testing, the error would have been caught and disaster averted. So things done up front can save a lot down the road.

Coverage is the final defense against software errors. We like to call structural coverage an analysis rather than a line of defense. It indicates the degree to which you're done, then when you're done, and covers every function. Each element of code is proven to perform the functions properly.

For Level C, there are statements to be covered. For Level B, decisions need to be covered in addition to statements. For Level A, statement, decisions and modified condition decisions. It's the measure of completeness. It feeds back additional requirements or tests if everything is complete.

Testing within DO-178B and DO-254, therefore, is the most singularly different aspect of all the life cycle phases compared to other standards. It will easily consume 30-45% of your overall developmental budget. When performed as described in this chapter, you're well on the way to minimizing cost and schedule impact of avionics testing, and keeping one step ahead of your competitors---all while delivering a safer product.

# 18. Structural Coverage Analysis

Structural coverage analysis deals with the extent that verification satisfies all objectives, including comprehensive requirements tests. It also identifies certain aspects about code, coders and testers. There are characteristics it discovers during the coding activity such as dead code. Dead code becomes obvious because, no matter what is tested, you cannot get to that piece of code. You realize it's because it was written, and possibly used at one time, but when modifications came about it was no longer necessary and wasn't taken out. More functions were added to obsolete that piece of software and so it's identified this way.

**Structural Coverage Testing-1**

The final defense against software errors

The most tedious and expensive form of protection ever conceived

Structural coverage; a "measure", not a "test"

"Coverage"; the extent to which a verification activity satisfies its objectives

**DO-178B calls for two types of coverage:**

- Requirements coverage
- Software structural coverage

**Why structural coverage?**

- Prove code was covered to degree required for criticality level
- Prove absence of unintended functions
- Establish thoroughness of requirements-based testing

S136/137

Structural coverage analysis is not done for Level D, but is required in increasing doses at Level C and above. You may find that the coder was a little overzealous and started adding functionality without requirements for that functionality. He said, "It'll be good, because, in two years, I'm sure they'll want this." Things that won't get tested, won't get covered.

These are several key reasons to perform structural coverage. There are additional requirements for Level A, and only Level A software, that you cannot do only structural coverage at the source listing level. You must also prove a correlation between the source and object, which means, in addition to everything else you did for Level A, you prove that object code does not have additional paths not obvious in the source listing because the compiler introduced it. That's why using compiler optimization at Level A is problematic and will add additional work for you to validate that optimization.

## Structural Coverage Testing-2

**The intent of structural coverage is to:**

Provide evidence that the code structure was verified to the degree required for the applicable software level

Provide a means to support demonstration of absence of unintended functions

Establish thoroughness of requirements-based testing. Structural coverage analysis provides a means to confirm that requirements-based tests exercised the code structure

*(from FAQ #43, DO-248Av14 FAA National Software Conference-June 6, 2001)*

**Structural Coverage is a white-box test, e.g. open the code**

**Required for Level C & Above:**

Level C: Statement coverage
Level B: Statement & decision-condition coverage
Level A: Statement, decision-condition, and modified condition decision coverage (MCDC)

**Source code or binary?**

S138/139

Because of the combination of constructs in the compiler and source code, additional elements of software may be introduced by the compiler. So you have to prove they are valid. If there are such elements, you add more tests to cover those additional object-level pieces. It's more difficult to do, with fewer tools available at the object level. Most coverage analysis tools such as VectorCast give you a screen view or printout of highlighted red, green and blue color indications for coverage analysis results. But they don't do it at the object level. So there is an additional challenge for Level A.

Normally, structural coverage testing is done at the white box level, sometimes even down at a module level for hard to reach structures. You see that Level C requires every statement, Level B, every decision. Level A, decision, conditions and statements, as well as modified condition decision. And the source to binary correlation is only for Level A software. So the source to binary requirement only applies to Level A.

## Statement Coverage

Proving that each source statement is covered and performs as expected

What is a "statement"?
Answer: smallest compilable entity, e.g. one line of code

**Decision Condition Coverage**

Proving that each decision has taken all possible outcomes

Also known as "branch" testing

S140/141

Consider, next, what we have to know while doing Level C or above. Remember what happens to criticality level; it typically only goes up, so eventually you may hit Level B and A for your software that was originally developed to Level C. Why does the level increase? The system becomes more important, pilot workload increases, redundancy is reduced, etc. What happens with structural coverage is that different degrees are required for criticality level. Recall that the largest cost delta among criticality levels occurred between D and C. Structural coverage analysis is the reason. That's unique to DO-178B. In fact, the document is really about determinism—proving innocence to show we're not guilty and having a strong process in general compliance with DO-

178B. It doesn't have to be exactly as described within 178B/254 because those are general guidelines, not process documents. But your process has to be at least as good while meeting those "guidelines". Most important, you must show it for your system's criticality level to achieve certification.

That's what is unique about DO-178B. It's the only standard with multiple criticality levels. For the Food and Drug Administration, medical instruments are Class 1 or Class 2 devices. But there isn't as large a difference between them. In DO-178B there is, which raises interesting questions.

Why don't we just make everything Level A? Would it be higher quality? Yes, it would. Why don't we do it? Remember who wrote the book; mainly industry. And their top three concerns are profit, profit and profit. Safety, certainly. But not safety at any cost, but safety at reasonable cost. DO-178B and DO-254 are reasonable. The good news is that if we do it right, Level A should only cost 25 to 40 percent more than Level D. Frankly, Level D compliance should be very easy to achieve because there is nothing required within Level D that any decent company would not perform anyway.

As consultants, we've gone to capable clients who were producing the first version of their product. Can you estimate the average version number on the software source code file? It turns out to be between 20 and 30. There are several versions and updates by the developer, then more with unit tests and a peer review. This is followed by many changes via informal and formal tests. Now we're up to about eight—but 20 to 30! What's happening? They are simply designing on the fly, which isn't designing, is it? It's guessing. Can you imagine building a house with your friends present and a pile of lumber? Just keep moving the foundation and walls around for a few years while you try different versions, right? Of course not, but then why is this very scenario engaged in by many in the software industry?

We've been to source code peer reviews where there were no completed requirements and traceability. So what was being reviewed? Is it elegant, beautiful, does it look like it might compile? Could I understand it?---all reasonable questions. But what was it supposed to do and what requirement was it trying to accomplish? Clearly, the code review was false and invalid. How can you review code if you've no real idea of what it is supposed to do? That's what complete and independently reviewed, requirements, in advance of any coding, are for. What does DO-178B say about a code peer review where complete and traced requirements are absent? Null and void, obviously. It simply doesn't count and you've wasted time.

A mid-sized company in Europe that builds avionics requested that we do an audit. We found almost no code review mistakes. It was a Level B project, done with excellent engineering, perhaps some of the better software we've encountered across hundreds of projects. The challenge they were facing was in reverse engineering. We asked, "How did you do the code reviews?" They were the smartest engineers we've seen in a long time and proud of their work. They sat around a table and analyzed everything. They didn't have all the requirements written down, nor written code review criteria, but they knew the requirements and their system very well.

How much credit could they get for those code reviews? Zero. How do you know on a checklist three or four pages long for each module that you checked four out-of-range boundary values on every item and every line of code? You don't know. Just because there are great checklists and code reviews are you going to improve the software that much? No. It's a lot more---it's the whole package. Avionics software is only as strong as its weakest link. DO-178B is about building a consistently strong chain through the entire engineering life cycle for the given level of criticality. Is a Level B chain as "strong" as a Level A chain? Of course not, but if the chain only needs to be Level B, why make it stronger? Exactly.

That's what the DO-178B life cycle is all about. Build quality by consistently following all processes and verifying them via the correctness process which includes testing. Build good compliance. With DO-178B, it's integrating, putting it all together and, what really makes it different is having structural coverage for each level of criticality.

At Level D there is no structural coverage. We're let off easy. At Level C every statement must be covered. It means we prove that each source statement performs as expected. The question is, "What is a statement?" DO-178B doesn't deal with that. How many lines of code are there? It doesn't say that, either. How do we count lines of code? Fifteen people count them 16 different ways!

A statement is the smallest compilable entity, usually denoted by a semicolon. At Level B, more must be done since we include all decision condition constructs ("branches"), proving we've executed them with *apriori* test criteria. For Level C we "only" cover every statement. Now we'll be 10 times better than most companies, but a long way from checking all the other structures. Why? It doesn't check branches and other possible decision conditions.

For Level B, it's all aggregate and cumulative. Statement coverage is done plus decision condition coverage, which is a fancy word for branch coverage; testing both sides of the case statements, switch statements and loops. We have to prove that each decision accounts for all possible outcomes. It's often informally referred to as branch testing.

Let's see how to cut structural coverage costs in half. Normal range testing, robustness testing and functional testing are all included within the DO-178B verification process. Functional testing comes first for us, which is requirements testing. If there are detailed requirements, we pick up almost all normal range test cases, many robustness tests and a lot of the structural coverage tests. The size of these is fairly similar. But if we have detailed requirements, whose size is not specified by DO-178B, how detailed do they have to be? It's very subjective.

How many requirements per line of code do we need for detailed requirements within a 20,000 line of code system; 800, 900, 1000? It's hard to say. But this is where DO-178B has many fail safes and double checks. A good rule of thumb used at HighRely and other major avionics development companies is at least one requirement for every 20 lines of code. It depends, however, on the type of system. For example, an algorithmic intensive function, say a Fourier transform, would have fewer requirements per line of code than the logic for a combination of discretes, which would have almost one requirement per line of code. But often we see normal avionics software with fewer than one requirement per line of code: this begs the obvious question, "Just what did the developer assume when writing the code?" And the DER/FAA asked the same questions, which had unpleasant results …

If there are no detailed requirements, you still have to do normal range and robustness testing. And write many more separate test cases for normal range and robustness tests because you didn't have them for the detailed requirements that you had in functional. So DO-178B catches you. Either way you're going to end up with similar tests and requirements. The only question is will you save time by doing them early in the project, or add them later at a much larger cost? When the DER or QA person starts looking, he asks, "Why do you have so many normal range and robustness test cases outside the realm of functional test cases?" The answer; "Because I didn't have detailed requirements as I should have earlier in the program so our QA person and DER made me add them as an exit criteria to the test phase."

The solution is to go back and add more detailed requirements before you start designing the software. If you add the requirements after you write the code (an interesting, albeit completely improper method of ensuring that code matches the requirements), then you might think that the ends justify the means and you are compliant with DO-178B. But this belies a failure in following your software development process and the QA person should prevent this. Even if your QA person also fails, the DER and FAA will be able to use your own configuration management system to determine the state of the requirements at the time of the code review. And those requirements had better be quite complete or your project will fail.

A book written about 40 years ago called "The Mythical Man Month" (which we still think is valid and a must-read for any engineering manager) compared software organizations to determine the ideal team size. The author concluded there was an ideal number; one person! Because engineers are better at quantifying than communication, the one-person team won't confuse itself by failing to communicate. But in the real world we must make assumptions, have detailed requirements and write them down so someone else can check them. You don't want the same person writing and testing the software. Even though that is legal for Level D, it is foolish.

If it's a Level B project, you need independence. At Level D you're legal without independence. Can you upgrade to Level B later if you haven't provided such independence? No. You must revisit everything. Why would you perform a review without doing it independently anyway? It's a little more expensive because someone else has to attain the learning curve. But reviewing your own work is simply foolish and the authors can attest to such from personal experience.

## MCDC Testing

**DO-178B Definition:**

*"Every decision has taken all possible outcomes at least once, and every condition in a decision is shown to independently affect that decision's outcome."*

**A condition independently affects a decision's outcome if that condition alone affects the outcome**

S142

That brings us to structural coverage. If a good job was done of writing detailed requirements and normal range and robustness tests, 70 to 80 percent of all source branches and structures should now be covered. True, you probably won't cover many MCDC conditions via functional and robustness testing, but most code does not have many MCDC's and those are only required to be covered for Level A anyway.

How expensive can it be? Many people estimate that Level A is 30-60 percent more expensive than Level B. That's common industry belief but, in reality, it is very excessive. Actually, Level A adds 5-10 percent to the cost over Level B. For Level A, you verify that every decision takes all possible outcomes at least once, and that every condition in each decision is shown to independently affect the outcome. That's right out of DO-178B. It means we have to tweak within a line of code each condition while holding all conditions constant, and proving the output changes as a result of tweaking that one condition. It's a good check for design, integration and logic and, in fact, it's almost a compiler check.

## Structural Coverage Example

### Using WOW (Weight on Wheels) and TR (Thrust Reversers)

```
Begin
    If WOW & TR activated:
        Activate_Thrust_Reverser;

End;
```

*Questions:*
*Level D structural test cases?*
*Level C?*
*Level B?*
*Level A?*

*S143*

Let's look an example of a function with several lines of code.

The end is just a compilation designation, which doesn't actually go in the code. If there is weight on wheels (WOW)---say, the airplane just landed---you want to be able to activate the thrust reverser. It shouldn't deploy before there is weight on wheels. And that's what the code looks like on the proximity sensing and thrust reverser control system. Well, not quite. There are over 10,000 lines of code to control this on the actual system. Thirty-five to 40,000 in a typical thrust reverser-proximity sensing system for a large commercial aircraft.

Our engineers developed and verified such software to Level A on the B-777. But only 20 percent needed to be Level A. To keep costs down, the rest was certified to Level B and C. This may sound unusual for such a critical system as a thrust reverser, which is often required for a safe landing, especially on short runways. The answer is that the system can be partitioned, showing that some code isn't as important as other code. It's an approach that can reduce costs by 15-20 percent. The idea is to do minimal work, but not compromise safety. That's what criticality levels are for. Again, it would be simple to do it all at Level A. Or, for that matter, certify all avionics software to Level A. But DO-178B and DO-254 are "reasonable" standards, largely developed by industry. Cost and schedule matter.

In the example above, you can complete the structural coverage analysis with one test case, showing that every statement is covered. There is one set of conditions that makes the If statement above true, such as weight on wheels positive and thrust reverser activated; then the line that activates the thrust reverser gets executed with one test case. This achieves Level C compliance.

There is a requirement in the specification for this situation. We would have achieved that structural coverage when we did functional tests. This is a normal operating condition for testing thrust reverser logic. If we missed it by not having it as a requirement, look at the code. We would write a normal range test case and if we missed it there, and did the third level of testing, robustness testing, we'd find it. So is it conceivable that we would have to write additional structural coverage test cases for Level C? No. We'd be truly lame to have not verified such a critical requirement and code block during functional and robustness testing, but in case that happens, that's what we have to do.

There are six levels of checks just for Level C. If we built a cell phone, for example, there would be significant amounts of logic and hundreds of thousands of lines of code. I would do at least a Level C process, in fact, for any commercial embedded non-avionics project to ensure quality. And that's why the medical device and nuclear power folks are adopting DO-178B style processes—not for Level A—but for Level C.

Next is Level B. You could probably do it in a minimum number of three test cases to be safe. But the minimum is two. All I need is one set of conditions where the "If" statement is false (not activated) and a repeat of the Level C test case, where the "If" statement is true. Then I've covered the branch because it's not complex. Complex refers to a set of conditions for which MCDC needs to be invoked.

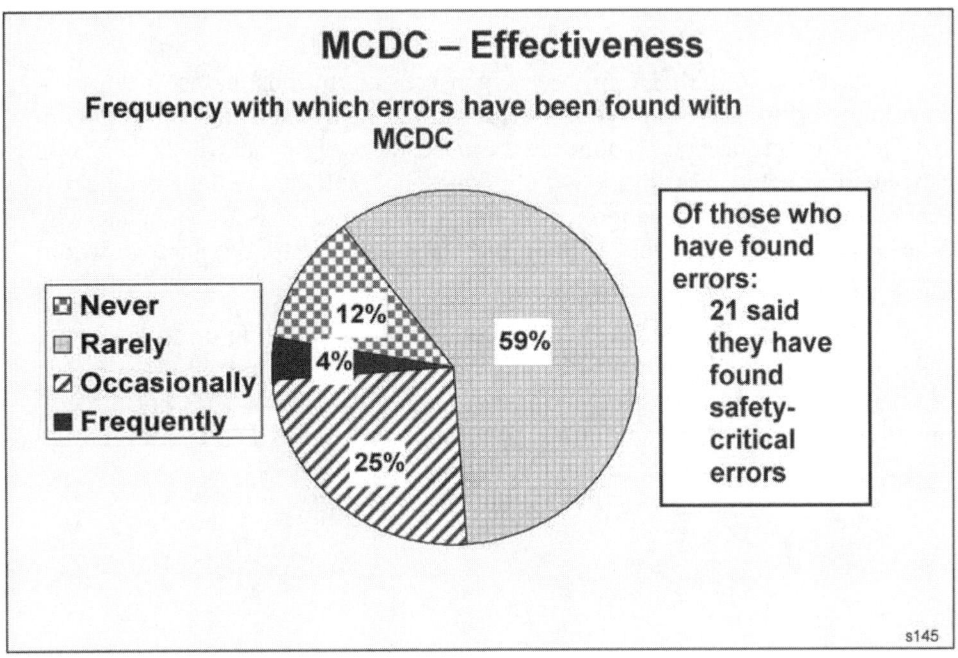

## Structural Coverage

Structural coverage, even MCDC, does not guarantee "quality"

MCDC has taken on 'urban legend' status for utopian testing

Does not directly improve quality; merely assesses quality and degree to which prior testing achieves code coverage

S146

Fifteen years ago, before we had DO-178 tools such as commercial structural coverage analysis tools, we printed out the source listing at 132 columns per page. With colored pencils (not pens) we drew a lot of circles and said, "Test Case 127B does Condition Y." Somebody would change the source code and you would do it again. Now we have tools to greatly reduce cost by automating this structural coverage analysis, the most tedious of DO-178B processes

In conclusion, MCDC is very expensive and, in most cases, it does not detect many errors. In our projects, we don't find many meaningful errors with MCDC. One airline crash happened because the weight on wheels test was changed incorrectly. Testing wasn't redone. The software failed to detect it because the software was changed. And that's what Level A is all about and why it's tested at the system level. If these were actual hardware outputs, we would have to test it on hardware.

MCDC usually doesn't produce much, but if I'm flying on it, I'd prefer MCDC being done. Even with MCDC, quality is not guaranteed. MCDC has become urban legend with great utopian testing. It's not. It does not directly improve quality; it assesses the rigor of prior testing and is just one link in the DO-178B chain to help us attain safer systems.

# 19. Test and Tools

Can we test everything on a PC if we cross compile to Motorola 68332, for example, or a power PC? No, we can't. In fact, you have to test the actual target binary for everything, even for Level D. But testing that binary on a PC or simulator is not valid if the processing or output pertains to the hardware. So, if it's timing-related, bit-related, stack-related, BSP- or operating system-related, outputs or discrete IO, we have to test on actual hardware.

But we get one little freebie. In algorithms, logic decisions and exception handling, we don't have to test on hardware, as long as we show the equivalency of the test platform to that of the target platform. There, we can test on the simulator. But the industry that wrote DO-178B doesn't like to destroy money, so if the test is destructive to RAM, for example, you don't have to test on actual hardware or actually crash it. Nothing has to be destroyed, even on destructive tests. But everything else, if it relates to hardware, must be tested on hardware.

There are many tools for testing. In choosing among them, each has positive and negative aspects.

**Tools, Pro & Con**

- Debuggers
- Emulators
- Simulators
- System Test Environments
- Unit Test Tools
- Test Case Generators
- I/O Stimulation/Capture (GPS, ARINC, 1553, etc)
- Structural Coverage Tools (including Instrumenters)

S149

Look at the tools, along with their pro's and con's. Everyone wants tool nirvana; that single testing tool that supports all their DO-178B needs. But we haven't found a single tool that lets you debug, test at the system level, test all requirements and analyze all structural coverage. We believe you're going to need at least two tools from the list below to accomplish your objectives.

Debuggers are wonderful; set the program counter to a line of code anywhere, propagate all conditions, single-step and get excellent structural coverage. But what is actually covered? Not much. You only prove the compiler compiled and the debugger worked. Most debuggers don't collect and aggregate structural coverage, and you won't get credit for hardware related testing. And most debuggers are weak in the automation and scripting of test cases department.

Emulators are great and provide a better job in evaluating hardware related operations but they typically are not automated. You usually cannot do much scripting with them. And you don't get full hardware credit on all the IO.

Instruction set simulators are powerful tools, with automation, batch capability and batch languages. You can simulate everything, especially the instruction simulators. But do you get credit for hardware IO? No. But you can connect some good structural coverage tools with them, and some even provide built in structural analysis capability.

System test environments are what DO-178B really likes; holistic testing. But there's a problem. Do they afford low-level intrusion to see all hard to reach conditions? Typically not. And you don't have automated structural coverage capability inherent in most system testing tools.

Unit testing is wonderful but it's difficult to obtain DO-178B credit for it since you're often not executing the software under test in an asynchronous, real-world environment. Stubs and drivers only go so far toward setting up realistic processing.

Test case generators help figure out which test cases to use. IO stimulation is needed if you're doing something like 1553 or GPS. And you really need structural coverage tools.

You don't want to do the manual source code printout and colored pencil approach to structural coverage if you have more than a thousand lines of code. But determine how to connect a structural coverage tool with another test execution tool to get credit for structural coverage. Strongly consider using a multi-purpose tool such as VectorCast to aid in your DO-178B testing and coverage analysis. Perform functional testing, normal range testing and robustness testing, and cover 80-85 percent of all possible branches. Then you do 10-15 percent of the remaining structures. Which 10 or 15 percent? The ones the tool says haven't been covered.

## Popular Software Testing Tools

**VectorCast**

Fastest growing, avionics specific. Very proven

**GCover**

Green Hills integrated tool with coverage analysis

**IPL**

Traditional coverage and one of the originals

**Logiscope**

Traditional coverage

**CodeTest**

Emulator based

**LDRA**

Modern analysis/coverage

**Rational**

Full suite, but less support after acquired by IBM

S150

What are some good tools? All those shown above have been successfully used on avionics projects. We typically use VectorCast and GCover. Other manufacturers make good tools qualifiable to DO-178B, but they all have pro and cons and different cost structures. Tool qualification is important. Because a tool's been proven to work in one case it doesn't mean you've used it completely and correctly so you still have to re-qualify the tool for each project. Refer to the tool qualification chapter in this book for more information, but remember there are two types: development tool qualification and test tool qualification. Development tools are tools whose output actually flies on the aircraft, and development tool qualification is quite difficult and expensive. Test tools are tools that evaluate other software which flies on the aircraft, and their qualification is usually not too difficult. But it helps greatly if you can procure (plan on paying) tool qualification data from the vendor and just re-execute the tool qual within your environment to show acceptability.

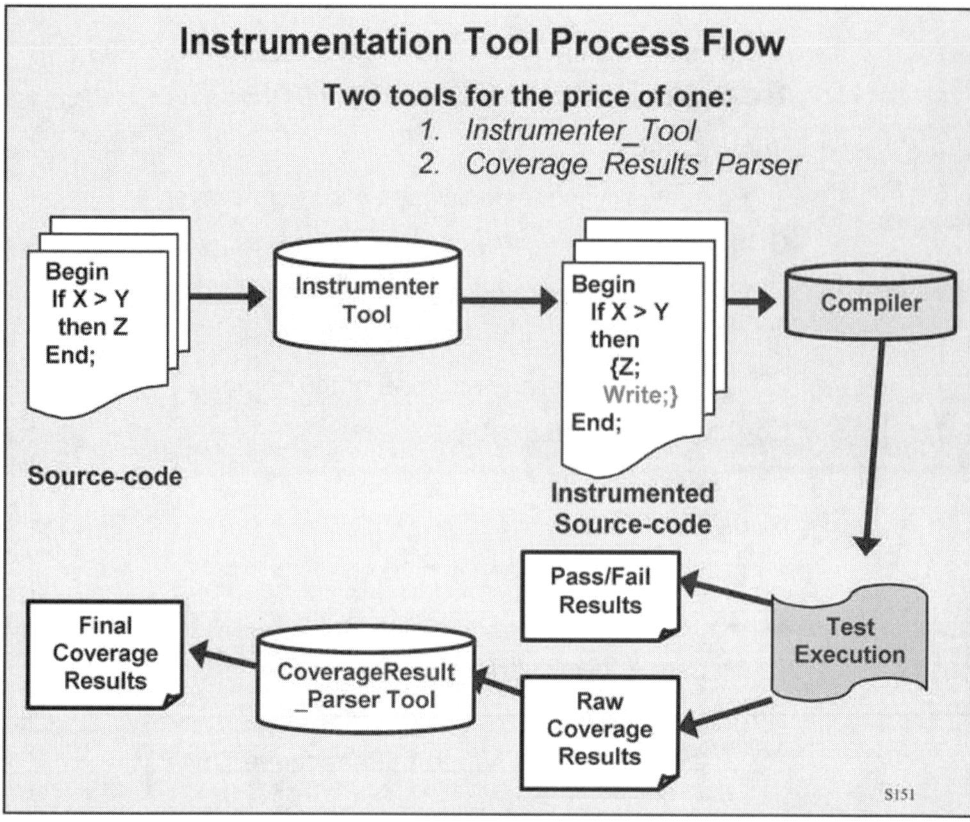


In automated code coverage a process called instrumentation is usually applied to the source code. The meaning of instrumentation, as shown above, is that you need two tools to perform coverage analysis, so structural coverage tool vendors supply you with two tools for the price of one. Note that the source code runs through the instrumenter, which adds special constructs to your source code, enabling coverage data to be recorded upon subsequent execution. You recompile, rerun the test, and check if all pass-fail results still pass where expected. Then, while executing your tests, the instrumented code that was added propagates special tables and coverage data via outputs from the tool to determine structural coverage. The second tool parses the recorded coverage results data generated by the first tool and provides various degrees of information regarding coverage. Tools like VectorCast show coverage results for each criticality level of DO-178B, and indicate what structures have been, and which remain to be, covered.

Remember, you can think of the "D" in DO-178B as meaning "determinism." What does the FAA think about a tool that changes the code via instrumentation, as described above? Are you going to run this instrumented code in your avionics box? The answer is "No". You won't have the necessary timing response, since instrumented code executes more slowly. You don't really trust it, and neither will the FAA. So, can you test this way? Clearly, such instrumentation tools must be approved in some fashion. Yes, hundreds of avionics projects have been certified. So what does that mean? Well, you can use it, but you have to run the test twice, which is twice as much work. The test must run once without the instrumenter to prove that all tests pass and the ones that fail are fixed. Then run it again to prove those same tests pass and fail, but get the structural coverage. Tests on timing may not pass the coverage results collection phase because everything slows down 20-30 percent. So now there's explaining to do because that's as good as it gets. The FAA is reasonable and so is DO-178B. If you can convince yourself that it failed only because of timing, you can document such failures accordingly and it should pass. Annotate it, being sure it's that condition. Have QA, a DER or an independent reviewer look at it and confirm the assessment.

Does instrumentation change the code? Yes it does, and you wouldn't want to leave instrumentation in your code. If you instrument, you must qualify the instrumenter to prove it's not doing other things.

Structural coverage requirements within DO-178B are undoubtedly the most costly additional work you'll incur to achieve compliance. Structural coverage can easily consume half your testing budget, with many companies actually spending more on coverage analysis and test cases than on the actual code development itself. But the trails have been blazed and modern tools now exist to assist you. So procure the right ones, learn how to effectively use them or outsource to a company that does, get early DER approval of your testing strategy and you're well on the road to successful certification.

# 20. DO-254 (Hardware)

If you'll work in the DO-178 software world, be aware that hardware people are asked to comply to a standard, as well. It was mostly imposed on people who program silicon-based logic (formerly called "firmware") via VHDL or any kind of HDL software. It's also required for those who do schematic capture. Anything that goes into an ASIC, FPGA, PLD, CPLD or put on silicon has to meet the standard. Recently, COTS microprocessors and controllers were allowed more leniency in complying with DO-254.

DO-254 was released in 2000 and generally required by the FAA beginning in 2005. Europeans have their commission---EUROCAE---to sanctify these guidelines and call it ED-80, their version of DO-178B, which has the same content as ED-12B.

**DO-254 Synopsis**

RTCA DO-254/EUROCAE ED-80, "Design Assurance Guidance for Airborne Electronic Hardware"

Released in 2000

Formally required by the FAA in 2005

Provides guidance for the design assurance of complex electronic hardware (CEH) in airborne systems and equipment

Counterpart to the well-established software standard RTCA DO-178B /EUROCAE ED-12B

s243

Even though DO-254 existed in 2000 the FAA issued an Advisory Circular in 2005 instructing applicants with hardware (particularly Complex hardware) to use it. What happened between 2000 and 2005? There was heated discussion on how D0-254 should be applied. The document provides guidance for designer assurance of complex electronic hardware and airborne systems and equipment. Once again it relates to airborne, not ground, systems.

CEH---complex electronic hardware---is an interesting acronym. There were many discussions about what's complex and what's simple. Today we don't care because everything containing silicon-based logic is deemed complex electronic hardware. If you determine that hardware is simple, e.g. contains no silicon-based logic, or you can fully test it, then no further application of DO-254 is required. But the document doesn't help because it says that anything that's not simple is complex---not exactly a precise requirement.

Simple is determined by the ability to subject hardware to all possible input conditions and proving that exhaustive testing occurred. Then it's simple. Test a resistor, for example. It may be tough to test exhaustively but equivalence classes over the range of voltage and current values allow you to say you've done exhaustive testing.

Consider a 4-bit controller and its discrete IO; two outputs, two inputs. This could be tested exhaustively and called simple. Let's go to a 16-pin device; eight inputs, eight outputs. You can exhaustively test all potential inputs, show all potential outputs and call it simple. That's discrete logic. Next, go to a thousand gate or million gate FPGA with 400 pins. You cannot test it exhaustively. Any time an ASIC is used, the task won't be simple. Look at DO-254 and it's in the exact realm as DO-178B; there are system requirements, advisory material, development guidance, verification guidance and safety assessment guidance. The documents that emerge are the volume constraints, system aspects, hardware design life cycle, planning process, additional considerations, design assurance, Levels A and B, C and D, and Appendix B. It talks about derived requirements. DO-178B never really talks about derived requirements, and it was decided to address them in DO-254.

**DO-254 Scope**

Applies to complex hardware

If "simple," no further application of DO-254

But when is hardware simple vs. complex?

S198

It has a design process that's the same as 178B; supporting processes, V&V, verification, CM, quality assurance (termed "process assurance") and certification liaison. It looks much like 178B because many of the same people wrote it. The reason we started that committee was because software people said, "Let's move the functionality to firmware so we don't have to follow 178B if we call it hardware or firmware." The FAA put a stop to that and a committee was started. We put the process together, which looks like 178B, to control development of VHDL, HDL and anything put onto silicon-based logic.

Some members of the committee said, "Wait a minute. This is hardware guidance. Let's put all circuitry into it---LRUs, microprocessors, cables and discrete devices, any combination of those circuits, CCAs and circuit card assemblies." Between 2000 and 2005 the committee argued over whether they were applying the document to hardware, firmware or software.

In 2005 the FAA decided that DO-254 should focus on the (complex) software portion of the hardware." This was done because in 2002 someone applied DO-254 to a CPU on a large transport aircraft and triggered a lot of concerns because of not meeting guidance for hardware with a CPU. It was an expensive two-year battle; when it was finally decided that CPUs are exempt because they're COTS silicon, have been in existence for a long time and are acceptable as a known quantity. (Not to mention that it would have been generally impossible to obtain manufacturing data from the vendor in support of DO-254!)

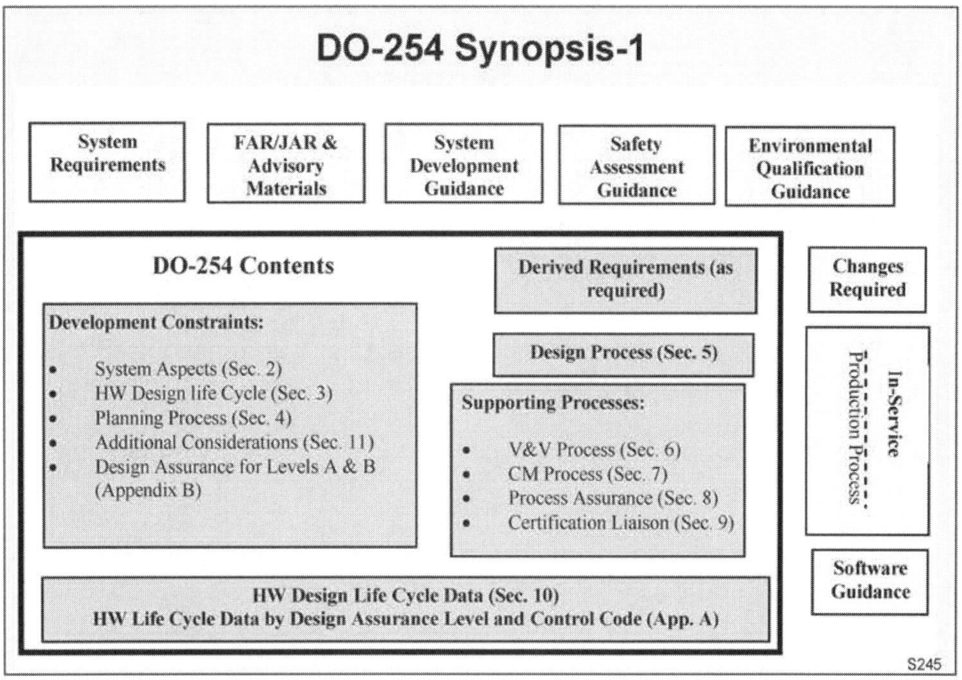

DO-254 talks about FFP (functional failure path), a safety assessment process at the hardware level where you trace paths through the silicon and architectural components to find and partition critical components. Can these components be partitioned by functional failure path analysis---or more like an FTA, or fault tree analysis?

You get system FFPs and equipment of FFPs so an LRU will get an FFP. Finally, decompose the FFPs where criticality is at a circuit card assembly. Eventually you get to the component level, then inside it. Let's say it's an ASIC with multiple blocks and the FFP says the critical piece is one block. It's Level A for one piece and Level B for another. This other piece is Level C, which doesn't contribute to a hazardous condition. So basically it's doing a fault tree analysis through the hardware.

## DO-254 Synopsis-2

Functions are allocated either as software or hardware and will be governed by the application of processes provided by the guidance of DO-254 or DO-178B respectively

*No more "firmware"*

DO-254 used as a standard in non-FAA projects; being recognized by military systems and equipment manufacturers

Not prescriptive or preferred life-cycle model, nor imply a structure for the organization

*Requires a hardware planning process, design, processes and supporting processes.*

s248

You do not need to do an FFP in hardware. DO-254 doesn't require FFP, which is a methodology. It's in Appendix B of the document; if a piece of hardware is determined to be Level A or Level B you'll have to do an Appendix B and everything else inside of DO-254. The FFP is required, just like a safety assessment at the software level, to partition something. The FFP is only required in hardware if you've got your FPGA and multiple components to separate out; one component to a lower level, another to a higher level. Then you do an FFP. Typically, most people don't do that, but say if it's in this PLD it's all one level.

It's also easiest to do because a lot of time can be spent doing FFPs only to find out everything's going to be level.

Guidance in the document addresses line replaceable units, circuit cards, custom micro-coated components, integrated hybrids, multi-chips and commercial off-the-shelf devices (COTS). DO-178B doesn't address COTS very well, with only a limited reference in Section 12. The document tells how to certify COTS products and how to make the argument. In 2005, Advisory Circular 21-152 was issued with FAA guidance that said forget everything else in the hardware; only do complex electronic hardware; ASICs, PLDs, FPGAs or anything on silicon. Don't do anything else about DO-254. Just concentrate on ASICs and PLDs. That's software-firmware process control.

## DO-254 Synopsis-3

**Guidance provides design assurance for:**

Line Replaceable Units (LRUs)

Circuit card/board assemblies (CCAs)

Custom micro-coded components (e.g. ASIC, PLD, FPGA, CPLD)

Integrated hybrids and multi-chip components

Commercial-Off-The-Shelf (COTS) devices

AC 20-152 concerns itself only with custom micro-coded components such as ASICs and PLDs

s247

DO-254 defines the system more clearly than 178B for allocation of requirements. It specifically says allocate requirements either to software or hardware, the concept being to get rid of software people putting their stuff into hardware, calling it firmware and avoiding a design assurance process.

DO-254 is also a standard in non-FAA projects. It's seriously being recognized by the military, which uses many ASICs and PLDs. They've not had a design assurance process.

Let' assume you have an FPGA with a CPU on it. Did you design it or was it purchased as an IP core? If you have design data, you apply DO-254. If it's an IP core you still apply DO-254. The problem is, you probably don't have the data. If you have a Xilinx FPGA and buy a CPU, you can obtain a CPU soft core and post it. You apply DO-254, doing much the same as in DO-178B; design information, traceability and testing all pieces. You can't just say because the FPGA is working with a DSP that the FPGA must be right.

What you've done is an element analysis, so you actually do the analysis of your tests. Trace the tests through all components on the FPGA and, if you can show that all inputs and outputs were exercised, as well as components on the FPGA, you've done the required element analysis. You still must have all the data. Software testing can be used to drive the element analysis.

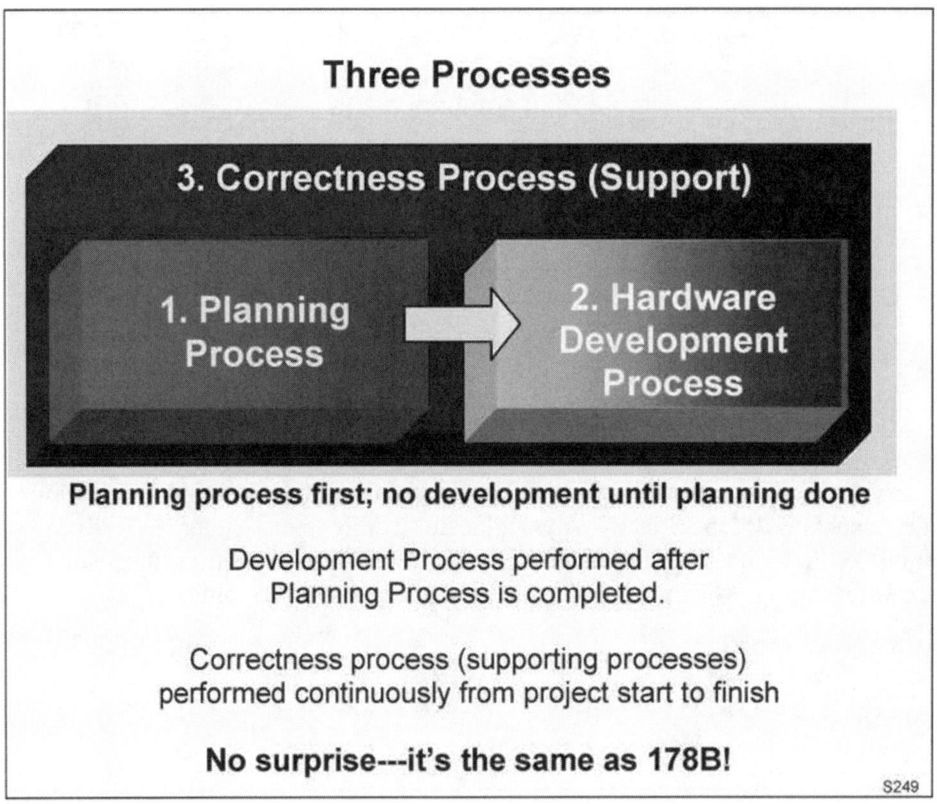

Like 178B, DO-254 is non-prescriptive. There's no preferred life cycle. It's a model. There's no implied structure or organization. The documents are developed as in 178B. There are three key processes; a correctness process, a planning process and hardware development process. The planning process must come first. Development follows planning, while the correctness process goes on continuously.

Here's the basic life cycle, similar to the waterfall life cycle presented in DO-178B. Hardware people have the same waterfall or sequential-type life cycle, although terminology is slightly different. Do all supporting processes and planning, but in an actual life cycle you do requirements capture. That's the same as the requirements definition in 178B. Do requirements, design and traceability identify those requirements? You do a conceptual design. It's a high-level, block-level diagram to show how the design looks in the architecture. You might call it high-level design or high-level requirements.

Detail design is determining how each circuit or component block works, how blocks interact, what traces from one block to another, what data gets passed, and what are the inputs and outputs.

Next there is implementation. It's the same as in the software world; first implementation then embedding within silicon. For hardware, there is coding and synthesis followed by the actual placement route. That places it on the silicon chip. There is a process in DO-254 that is not in 178; production transition, where you get hardware out onto the production line. So there's a process that asks if you are certain that what is designed is, in fact, what ends up on the production line. How is that process ensured?

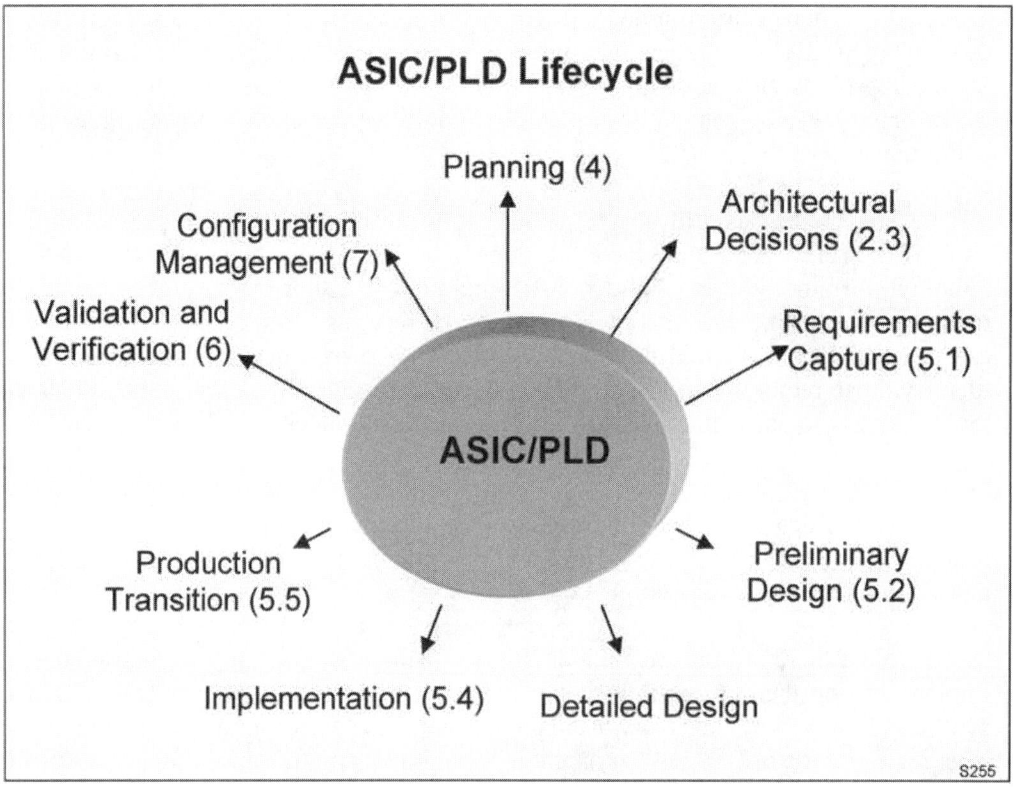


DO-254 data items look a lot like software data items. PHAC is the same as PSAC; production assurance is similar to quality assurance. The design plan, validation and verification plan are also similar. The validation planning is unique. In hardware design there are more derived requirements. On choosing a chipset, the choice imposes many hardware requirements; timing, interfaces and memory constraints, for example. You are required to have a validation plan specifically for derived requirements based on selected hardware. Be sure to develop derived requirements and validate the ones appropriate to the system. It is hardly practical to choose a chipset that doesn't have an RS232 communication port when talking to an RS232 system. Validate such requirements.

## DO-254 Data Items

***Plan for Hardware Aspects of Certification (PHAC)**

Hardware Design Plan

Hardware Validation Plan

***Hardware Verification Plan (HVP)**

Hardware Configuration Management Plan

Hardware Process Assurance Plan

Hardware Design Standards

Requirements Standards

Validation and Verification Standards

Hardware Archive Standards

Hardware Requirements

Conceptual Design Data

Detailed Design Data

***Top-Level Drawing**

Assembly Drawings

Installation Control Drawings

Hardware/Software Interface Data

Hardware Traceability Data

Hardware Review and Analysis Procedures

Hardware Review and Analysis Results

Hardware Test Procedures

Hardware Test Results

Hardware Acceptance Test Criteria

Problem Reports

Hardware Configuration Management Records

Hardware Process Assurance Records

***Hardware Accomplishment Summary (HAS)**

(Total: 27)

*Must be submitted to FAA

s251

Let's compare the number of specific documents in 178B with those in 254. By our count there are twenty-two unique documents or data items in 178B. Notice in the figure above that four of the twenty-seven DO-254 data items (shown underlined) are to be submitted to the FAA: PHAC (like the PSAC); the HAS, hardware accomplishments summary (like the software accomplishments summary); top level drawing; and the hardware verification plan.

The hardware verification plan is unique; the FAA wants to see it. The reason is that it's unlike 178B because there isn't a simple method for determining the metric of how the system was tested. The FAA wants to see the kind of analysis you're going to do, how it's applied, the analysis, how it's accomplished and what testing will determine if hardware is functioning to requirements. They also want to know what is missing from this document list. It's one of the required ones from software but not in the hardware list.

It made sense if everything is handled in hardware. In the old method you'd get a top-level drawing that covered the hardware. Finally, you get to the subassembly listing where it says you've got this or that PLD and it would be programmed on a PLD. You can usually find that information in the top-level drawing. What are we putting here, though? We're essentially putting software on the PLDs. We want to know about those pieces of software and their versions. What's all the configuration information?

Even though DO-254 doesn't call it out, this has to be satisfied for ASICs and PLDs, so you'll end up writing a software configuration index. Substitute a configuration index for the ASIC top-level drawing but still give them a top-level drawing. You must be able to find the ASIC, CCA or card it's put on, You need the CI, configuration index, that goes with the ASIC. This can be difficult because DO-254 does not tell exactly how to deal with an ASIC or PLD. You have to do planning, safety assessment, requirements capture and conceptual design, then a detailed design to the implementation. While these are being done, you do validation verification, configuration management and, at the end, production transition.

Most engineers who deal with VHDL and HDL or do some kind of silicon programming, basically following DO-178B methodology, which is identical to DO-254 except that it's not there! You can't construct a coverage analysis on silicon. There's no way to instrument the silicon. You can't get inside the little component you defined and see if you've covered all the paths. There are no tool sets. There is, however, Appendix B which covers advanced analysis techniques. If you're at Level A or B there is something equivalent to structural coverage analysis, which is called element analysis. It's a process of analyzing all inputs and outputs of a component and tracing via the circuitry inside the FPGA or PLD. With discrete logic, it's simple because you can do truth tables and, for every trace. If there is also analog and bus data it becomes more difficult, but you have to do element analysis. The good news is that you pick how low to go to with those components

You no longer have to consider LRUs or the circuit assembly. If criticality is less than Level D, DO-254 doesn't apply, so use the old methodologies. The Advisory Circular says commercial off-the-shelf microprocessors are not considered from a DO-254 perspective. They only matter from a historical perspective---do they have a "pedigree"? Don't try to use a CPU that does not have a solid pedigree in an airborne system.

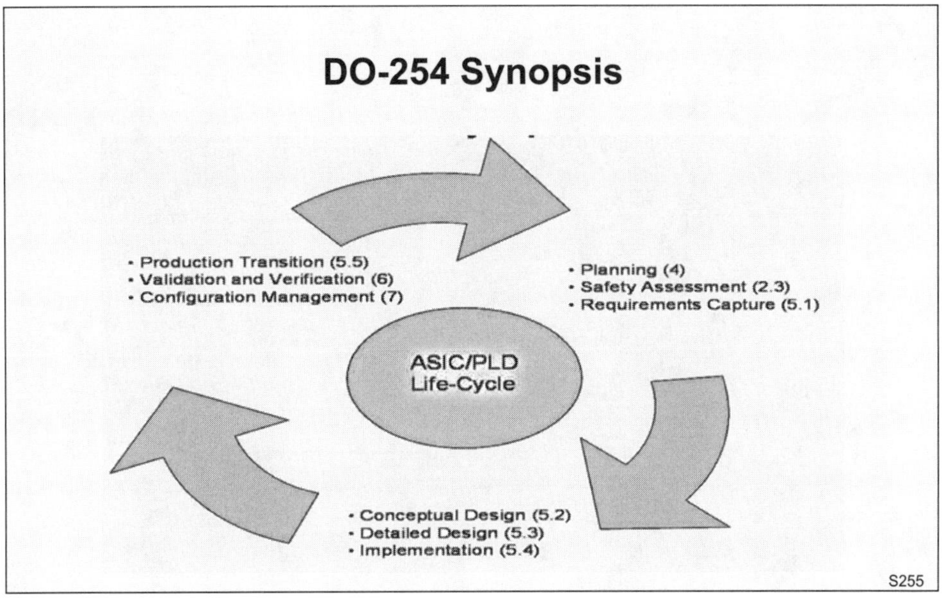


Note in the figure above there are sections going around a circle. You do planning, architectural requirements, preliminary design, detailed design, implementation, validation, verification and configuration management.

DO-254 describes the same four levels of criticality; Level A through D. According to the Advisory Circular, for Level D use standard practices done before. There are a top-level diagram and testing criteria. Tracing is done to be certain the pieces are correct. So Level D is out of DO-254 and Level A and B are identical. Even though there are criticality levels of A or B, the design assurance effort is identical and done the same way. That's why planning and design are the same for all four if you choose Level D.

Implementation is a little different. There's a higher workload for A and B, whereas in C and D there's little difference. The higher workload is element analysis, more reviews and lot more analysis for A and B. So one objective in DO-254 is to achieve Level C if you can.

For hardware planning, process objectives are much the same as software planning objectives. All hardware life cycle processes are defined, standards are selected and defined. There are four standards which add an extra standards document. In AC 20-152 there are design standards, requirement standards, V&V standards and archive standards. A fourth standard developed in DO-254 is archive standards.

# 21. Hardware Design Life Cycle

If you write a hardware requirements document or implement hardware requirements, you work through the same process. There are mission requirements. You need entry criteria, operational constraints, system-level specifications, engineering methodology selection and descriptions of the methods you are using. In functional analysis and allocation, you decide how to allocate the elements and functions of chips and components. The hardware is portioned into components, as done with software.

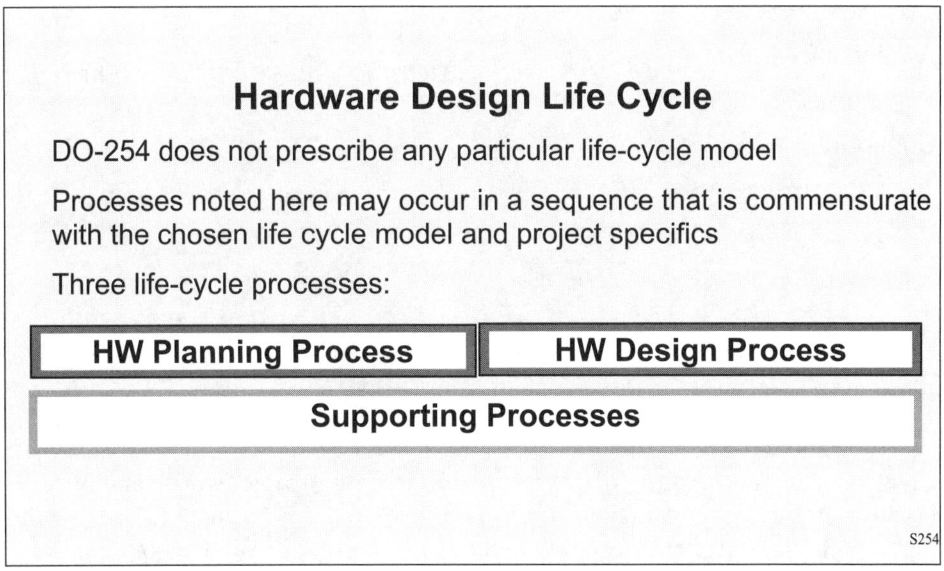

## Hardware Requirements

**System Architecture Development**

Alternative Systems Concepts Evaluation

System Architectural Design Verification

**System Analysis And Control**

Trade Studies

System Cost Effectiveness Analysis

Failure Modes and Effects Analysis

Availability Analysis

Performance Analysis/Performance

Configuration Management

Technical Reviews and Audits

Project Engineering Notebook

**System Life Cycle Management**

Life Cycle Items

Concept Stage

Concept Stage Outcomes

Development Stage

*Development Stage Purpose and Outcomes*

Production Stage

*Production Stage Purpose and Outcomes*

Utilization Stage

*Utilization Stage Purpose and Outcomes*

Support Stage

*Support Stage Purpose and Outcomes*

Retirement Stage

*Retirement Stage Purpose and Outcomes*

S259

Should avionics engineers worry about costs? If you're a hardware engineer you do because of trade-offs. Hardware people worry about costs because of the design decision of which chipset and components to use. If the run is high, that difference can be hundreds of thousands of dollars, so hardware engineers perform a cost/benefit trade-off.

System life cycle management relates to stages that hardware requirements might go through. You do a concept stage, a development stage, a production stage, utilization stage, support stage, then a requirement stage for hardware. There's a process at a higher abstraction level for hardware engineers because they have to worry about chipsets that might go out of production. So there is concern about more than just the development process.

## Hardware Requirements

### Introduction

Requirement Definition

Requirement Characteristics

Requirement Format

### Mission Requirements Analysis

Entry Criteria

Operational Concept

System Level Specification

System Interface Definition

Engineering Methodology Selection

### Functional Analysis And Allocation

Iteration/Leveling

Hardware Event Analysis

Partition Hardware into Components

Hardware Requirements Analysis

Create the Performance Model

Requirements Traceability

S258

## Hardware Design Processes

**DO-254 defines five major design processes:**

| | | |
|---|---|---|
| Requirements Capture | Hardware Requirements | 3 objectives |
| Conceptual Design | Conceptual Design Data | 3 objectives |
| | Detailed Design Data<br>Top-level Drawing | |
| Detailed Design | Assembly Drawings<br>Installation Control Drawings<br>Hardware/Software Interface Data | 3 objectives |
| Implementation | | 4 objectives |
| Production Transition | | 4 objectives |

DO-254 specifically addresses COTS. You have to indicate, as done in 178 and the PSAC, that you are going to implement COTS, and how it will be used. If there is a 1553 controller, and it's put in the CCA, DO-254 asks for its requirements, design information and traceability. What about reviews for everything sitting on the 1553 controller? With DO-254 you don't have to do them—just show that the controller is widely used and the manufacturer is known to have good engineering practices—or a "pedigree" and documentation.

On the hardware side there is an out. If it's COTS, and it's being used in cell phones, automobiles or the railroads—and the COTS chips have a good history by a reputable manufacturer, you're done. Obviously, it gets tested later when the system is tested and you still have to perform all that. So using product service history for credit is encouraged. It's worth finding a COTS product because it's a lot easier than trying to get at 1553 code inside of a chip.

If you'll use product history, check the manufacturer's quality control, established reliability of components and other such information often found on the data sheet. Here's an important point; if you'll be flying at forty thousand feet---and it's a critical PLD---don't get one that's plastic packaged. Metal packaging is recommended.

## Hardware Accomplishment Summary

**The PHAC told the certification authority what you were going to do. The Hardware Accomplishment Summary tells them what you actually did.**

| | |
|---|---|
| System Overview | Hardware Life Cycle Data |
| Hardware Overview | Previously Developed Hardware |
| Certification Considerations | Additional Considerations |
| Hardware Design Life Cycle Description | Alternative Methods |

This document must also identify differences from the approved PHAC and contain identification information for the hardware, change history and status of the hardware and, finally, a compliance statement.

*Note the similarity with the Software Accomplishment Summary*

S261

The hardware accomplishment summary tells the authority what was done on the hardware side. It looks like a software accomplishment summary; system overview, certification considerations, official considerations, hardware life cycle data, transition points, hardware and software overview, design life cycle description and the data you are creating. If you previously developed hardware, what COTS are you using, are there alternative methods for design assurance? In summary, there are comprehensive guidelines for design assurance in complex airborne hardware.

There are inconsistencies. DO-254 talks about hardware, not just ASICs and PLDs. When you try to apply DO-254 to everything, the FAA says, "You can't do that. We only want you to do ASICs and PLDs." Europeans, on the other hand (the old JAA, now EASA), often say "Apply all of DO-254 for commercial aircraft," although the Europeans seem be proceeding more slowly on requiring DO-254 for military aircraft. If it's an Airbus, or flies commercially in European airspace, you need to apply it, not only to ASICs and PLDs, but all hardware. Look at the CCA, LRU, circuits and connectors---all of it---and coordinate your PHAC in advance with the certification authorities to ensure you've made the correct interpretation in defining the scope of DO-254.

Before using DO-254 on a project be certain you have agreement of the certification authority on any final issues. It's a new standard and RTCA would also like to hear if there are issues it should know about. New documents need to eventually explain all of DO-254. There are regular meetings by the FAA and RTCA regarding clarification of DO-254; these are open to most DERs and some industry personnel. You can read the minutes and results of those meetings online.

There were sixty-six objectives in DO-178B. Those are the objectives in DO-254. They're quite vague but the committee decided it didn't want to make a checklist and not tell the user how to do it. The sixty-six objectives in 178B are the checklists in Appendix A. For DO-254 it was decided that you could do it that way, but still make these seventeen objectives. Subject design has three objectives, detailed design has three, implementation has four and production transition has four. It's a lot simpler to meet these objectives.

## DO-254 Summary

DO-254 provides a comprehensive set of guidelines for design assurance activities during development of complex electronic hardware

The standard loosely parallels DO-178B

As with any standard, particularly new ones, expect to find some inconsistencies and apparent omissions

Before using DO-254 on a project, ensure you have agreement with your certification authority for DO-254 application

Feed back your experiences with DO-254 to RTCA or the FAA so future revisions can be improved

S262

## Additional Sources


RTCA DO-254, *Design Assurance Guidance for Airborne Electronic Hardware* available at www.rtca.org

FAA AC 20-152, *RTCA, Inc., Document RTCA/DO-254, Design Assurance Guidance for Airborne Electronic Hardware*

*Complex Electronic Hardware Interactive Video and Self-Study Course* available at http://av-info.faa.gov/software/complexhdw.html

*Design, Test, and Certification Issues for Complex Integrated Circuits,* DOT/FAA/AR-95/31 Technical Report available at: Department of Commerce National Technical Information Service, 5285 Port Royal Road Springfield, VA 22161 (800) 553-6847

*An Introduction to VHDL, David Pellerin, Accolade Design Automation,* available at http://www.acc-eda.com/h_intro.html

*Application Specific Integrated Circuits*, Smith, Addison-Wesley Longman, 1997

RTCA DO-178B, *Software Considerations for Airborne Systems and Equipment)*

RTCA DO-160D, *Environmental Conditions and Test Procedures for Airborne Equipment*

ARP 4754, *Certification Considerations for Highly Integrated or Complex Aircraft Systems,* SAE, http://www.sae.org

ARP 4761, *Guidelines and Methods for Conducting the Safety Assessment Process on Civil Airborne Systems and Equipment*, SAE, http://www.sae.org

# 22. Gap Analysis

Now that you understand basic aspects of DO-178B and DO-254, it's time to consider how best to implement them within your organization. Gap analysis is the activity of comparing your current development processes with requirements within 178/254 and to ascertain differences, or "gaps." To understand gap analysis, let's begin by talking about gaps between DO-178B and your organization's processes. Depending on the application, the organization may have several gaps. If you're military in the United States you're probably using standard 498 or a 2167A derivative. Most development processes established in 498 may apply to 178B, but you want to know what's not applicable and what is missing. Find out where the gaps are and fill them. What we see, however, is that many companies don't know where the gaps are, let alone know how to do a gap analysis.

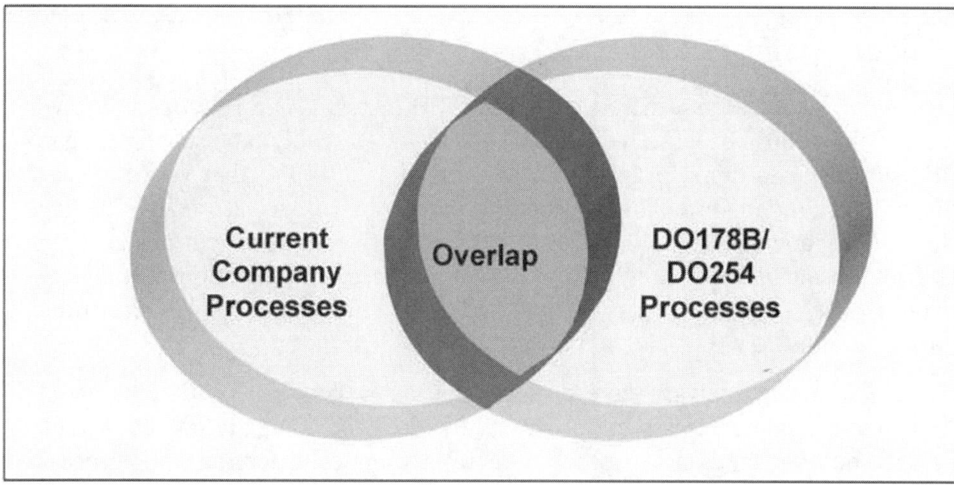


A common item in gap analysis for 498 is the VDD, your advanced version description document. That terminology originally came from 2167A and its predecessors. DO-178B provided the term Software Configuration Index (SCI). If you compare a version description document to the software configuration index, they're not exactly the same. There are gaps. A VDD specifically talks about software configuration and how to rebuild it. SCI goes beyond that. It talks about base lines you've maintained, problem reports put into that base line, software and documents that are part of the base line, along with what was rebuilt and tested. So the SCI is a large document. There's a gap. If you believe the VDD completely complies with SCI you'd be wrong, but a VDD combined with the results of the software development plan in 498 answers those questions.

DO-178B doesn't care how you package it, so long as all the data is there. If there's a 498 process, create a cross-reference table for what the SCI is and where the data is represented. It's a little more difficult but you don't have to write an SCI. Continue to use the VDD, but what's the gap? Gap analysis tries to establish that for you. In current company processes, DO-178B or DO-254, there's always an overlap.

## Gap Analysis

**Assess the "Gap" between your current processes/practices, and those necessary for DO-178B**

Typical Gap if no prior DO-178B history?

| | |
|---|---|
| SEI CMM Level 1: | 70% – 90% |
| SEI CMM Level 2: | 50% – 75% |
| SEI CMM Level 3: | 35% – 60% |
| SEI CMM Level 4: | 25% – 40% |
| SEI CMM Level 5: | 20% – 35% |

S268

In a typical gap there's no prior 178B history. As a rule of thumb, we find it's not 100% exact. If you're a Level 1, you probably have a 70-90% gap. It's surprising to think, even if you're a Level 1 at 100%, that you'd at least have 30% overlap---probably the coding piece. SEI Level 2, a 50-75% gap. It's coming down. SEI Level 3, a 35-60% gap. Even if you went to Level 5 in a comparison of what they're looking for in these organizations, even they would have a 25-35% gap. DO-178B looks for unique things that other processes don't ask for.

In a gap analysis, take a close look at every facet 178B talks about. Examine the safety assessment process; how does the company go about doing it? If you're a standard 498 process, there is a modest safety analysis process, usually done on the basis of mission safety or criticality, not from the survivability of the aircraft. Although there is a safety analysis process there are nuances—gaps that need to be modified and changed. They definitely don't assess it as Level A, B, C, D or E, but more of a one, two, three mission-critical aspect. And remember, 178B and 254 use the safety assessment process to iteratively refine system architecture to ensure your process accomplishes that, in addition to helping define the criticality level.

## Recap of DO-178B Processes

**Safety Assessment Process**

There are three system safety assessment processes: functional hazard assessment (FHA), preliminary system safety assessment (PSSA) and SSA.

**Planning Process**

The purpose of the planning process is to define the means by which the functional and airworthiness requirements are converted into a hardware item with an acceptable amount of evidence of assurance that the item will safely perform its intended functions.

**Architecture Planning & Development**

Given the safety, functional and performance requirements allocated to the software/hardware by the system process, the software/hardware safety assessment determines the software/hardware design assurance level for each function and contributes to determining the appropriate design assurance strategies to be used.

S269

Look at the planning process. Do you accomplish all five basic plans for the five basic processes? You probably do four, and do not do the fifth, the PSAC. Do you create the three basic standards? Most companies have some kind of standards for requirements, design and coding so you just need minor augmentation in these standards. If you're doing a 498 or DEF standard, British type, most data expected in the PSAC ends up in the software development plan.

## Gap Analysis

**Requirements Process**

Identifies and records the software/hardware item requirements.

**Conceptual Design Process**

Produces a high-level design concept that may be assessed to determine the potential for the resulting design implementation to meet the requirements.

**Detailed Design Process:**

Uses hardware item requirements and conceptual design data as the basis for detailed design.

S270

Look at your requirements process and perform an assessment to ascertain if your checklists meet the necessary criteria. An interesting exercise is determining if a requirement is really unambiguous. Look at all planning documents, safety assessment requirements, design code, functional structure testing, CM and QA. What do you think the gap analysis is from our perspective? When we do a gap analysis, a DER examines the information and determines if you're compliant or not. It's an effective gap analysis, and filling in the gaps will probably bring you into compliance. "Probably" because another DER might say you're not compliant and, of course, *all* DER's are correct! If, in fact, you obtain an opinion from your DER which you find difficult to accept, feel free to ask for case histories of companies with a similar situation. And you can always get a second DER opinion.

Typically, if you're a 2167, 498, DEF-stnds 55 or 56 type company, you most likely are between SEI Level 2 and 3 because the government says you shall be Level 3 if you want to do business with us. Note, though, that most Level 3 and 4 companies have processes that comply with that respective level. However, their in-practice implementations are usually one, and sometimes two, levels beneath that. In other words, they have nice processes but don't follow them or, if they do, can't prove such. Note that the PSAC usually has a large---about 80%---gap because most companies don't have the information required in a PSAC, which is a unique document required for DO-178B. Most other standards don't have anything like a PSAC, either. The QA plan has a small gap; the CM plan, a very small gap, if any at all. The development plan, typically, is also a small-to-medium gap.

Most companies have a development plan which may not answer all questions DO-178B looks for. The software verification plan typically has a higher, older-type gap because any software verification plan you're talking about typically does not contain structural coverage analysis processes and how you go about implementing them. Typically, in most other organizations structural coverage analysis is a unique aspect of 178B. You're going to find a large gap there.

## What should a Certification Gap Analysis Examine?

**Planning Documents (PSAC, CMP, QAP, SDP, SVP)**

**Development and Correctness processes including:**

- Safety Assessments
- Requirements, Design, Code
- Functional and Structural Testing
- CM and QA
- Tool Qualification
- Checklists, Reviews, Audits
- DER Activities

s271

Most companies don't do a safety assessment from the perspective of aircraft functionality. It's unique to 178B and most commercial organizations don't perform structural coverage because they see very little benefit compared to the cost. DO-178B is probably the only standard that calls for qualification as part of the development and verification processes.

Most companies have checklists, but not necessarily the ones required from a 178B perspective so that is a large gap. Another huge gap is DER liaison, which is a unique step for the FAA. Without somebody saying you are compliant you can't get certified.

DO-178B states clearly that a specific person who has knowledge should be doing the complete activity, with no TBDs. Correctness means there are no TBDs. We've received requirements documents for review and found large numbers of TBDs, which is not acceptable. A current practice to avoid TBDs is to move them all to star-star-star....or using blanks with parentheses. That won't work; you've got to fill in all the answers, otherwise the review is not complete. And until the review is complete, you cannot meet DO-178B transition criteria and move on to the next step.

A good answer in some cases is that you don't know what the requirement is, say, whether it's 10 or 100 milliseconds. Put something in as a requirement, then state, for example, that the 50 millisecond requirement is temporary. Don't leave a blank because the software engineer will fill it in since he may have to finish the software tomorrow. He will determine what it's going to be and inserts the number. Three months later, the number is still there because no-one checked. Now the number causes a problem with lag time that nobody down the chain analyzed. The software is flying and management says, "Bobby, we're three versions down. This is good." Guess what happens when the engine management group says, "We've got to test all over again because 10 milliseconds was used since that was the requirement at the time. It should have been 100."

So TBDs shouldn't be there except for very rare cases; about as rare as a To statement. So, define real requirements and be sure to return to recheck later in the design process, which asks if you resolved all TBD's. A good means to ensure you do that is to generate a software change request and keep it open until that issue is resolved. It assures you won't forget about it later.

Be sure to keep terminology consistent. If you have a term "high-level requirement" in one document, don't say "conceptual" somewhere else. DO-178B and 254 are about consistency in your process, so you must be consistent in documentation terminology. If something is "green" in one document, it'd better be green in another. If a dollar sign is the entry for a specific variable, use it all the way through the system.

Traceability obviously traces system level requirements down to high level, then low level hardware/software requirements, and back up from code and tests. What if the system doesn't have system level requirements? In one actual case, a group of smart engineers developed a commercial product (that became very successful), using requirements that existed only in their heads. But because of the demands of DO-178B, the product had to be reverse-engineered to obtain system requirements It negatively affected the design, which had to be updated at great cost and market delay.

The reality is that you will find these requirements in the real world; they exist in your system requirements. Does that mean you have to do it as systems integrators and systems developers? You write systems requirements a certain way so they're more quantifiable. You may define accuracy as plus or minus twenty feet in the horizontal, and plus or minus ten in the vertical, which is a more quantitative, more accurate, requirement. The system may respond to safety-critical pilot actions in less than 40 milliseconds; non-safety critical pilot actions in less than 100 milliseconds. We decomposed this one into two requirements. If I'm re-writing my systems spec am I decomposing? I'm just re-writing my systems spec because somebody might have written poor or unclear requirements.

What usually happens in real life is that the customer defines the basic features and your sales team further elucidates them. You might get an operational concept document or procurement spec, then a systems spec, or you'll have to write it yourself. Writing a systems spec for an operating system is mind-boggling but you iterate and iterate until there is a good system specification.

Consider non-testable requirements. DO-178B calls for robust software, for example, which shall degrade gracefully under stress. This could mean to wait 10 seconds and provide a warning before you shut down, restart and enable the back-up system to take over. That's one definition of graceful.

DO-178 requires software to be developed in accordance with modern programming practices. What is modern?

"Software shall provide the necessary processing under all modes of operation." What are the modes? Do I make up the modes? What is the necessary processing? In a display system we have a VMC mode, IFR mode, a coupled mode, a navigation mode and joystick mode; which shall be covered to what degree by requirements and implementation?

Computer memory utilization shall be minimized to accommodate future growth. This could also be defined as "software shall be easy to use". You've seen that in every requirement doc or customer spec ever received; software shall be easy to use. Most of the time the customer touches only the system, not the software.

The problem is in unquantifiable words such as; sufficient, modular, achievement, adequate, proficient, accomplished, possible, better, higher or slower. They have an infinite number of meanings to an infinite number of people. That's not engineering. We want consistency, accuracy and precision.

Where does it fall apart? If it's a requirement, why is it a goal? That's a consistency problem. A requirement means you will do this—it must be done—it's not a goal. A goal means you *want* to achieve it. It shouldn't have been a requirement, and falls apart on the "sufficiently" accurate aspect. You can't quantify "sufficient". If there are requirements and system specs, write requirements this way. Get rid of unclear requirements and specify good ones.

If you have low level knowledge, design details creep up and you call them requirements. Next you'll be asked for detailed requirements. If all design details are up in front, somebody will ask you to create detailed requirements, so what do you do next? You write code for detailed requirements, then write tests for all lines of code, specifically because they became your requirements. It becomes a mind-boggling job.

If you have in the systems spec or high level requirements all those details, stop. If someone asks, "Where are your detailed requirements?" say, "Those are my detailed requirements." It's justified. You can code from that.

**Data flow inconsistence.** There was a time when software development data flow diagramming was the way to write software requirement specs. You would write around a data flow diagram and describe what's happening. Some people would then change names on the data flow or use a different prompt for saying the data flow went somewhere else. Avoid those inconsistencies by always being specific.

**Desire to have outputs not derived from the given inputs.** If you write bad requirements you will get that result or worse. All scenarios were not identified and thought through. We like the UML use cases because they force you to go through scenarios. It's a good way to elicit requirements. People want to use cases as a description of requirements, but it's problematic because it's a method of elicitation. You have to take elicitation from use cases and design them as requirements. Engineers don't specify errors they know tightly. You shouldn't have an engineer who understands the system extremely well write the requirements. If he completely understands the system he's not going to do a good job of writing requirements or specifying because he mentally assumes so much and fails to record it via specification. It's the opposite of what you would think. They assume other people know it as well and simply don't write down little details necessary to remove assumptions within other readers' minds.

Engineers generally specify only areas they know and understand. Ask at a meeting; "Do you know about this?" and everyone says "Yes." But start asking questions and the reverse is often true. We've seen requirements specifications with this situation; the person writing the requirements doesn't know much about it, so he writes some general specification (and keeps his job). So watch out for this situation. You really need a thorough requirements writing standard, process and checklist to ensure all requirements are written consistently and thoroughly.

Requirements are the foundation for the development. If you find a requirements writer and system engineer who know what they're doing, pay them more. Reduce the number of system requirements writers because it's better to have fewer good people than a larger number who commit ambiguity and other errors which cascade down software life cycles. Truly, requirements are your foundation and your success is established right up front.

Scrub your requirements regularly. It is not satisfactory from a DO-178B perspective to take requirements specifications and say you reviewed the document in a few minutes saying, "The QA guy and I reviewed that document and all requirements are valid." That no longer flies, literally.

The requirement needs to be reviewed, not the spec. There should be evidence of the review, not review of the spec. There's a distinction. Scrub them often. It's better to use a tool because you can do that scrub and track the review of the requirement.

If you use an SRS document process, you do a lot of scrubbing and the document will go rev, rev, rev—forever. That's a lot of wasted effort and paperwork.

Most software defects are due to poor requirements. We've known that for fifty years. Every book that's ever been written by top guns of software in the education world says it. Every study says, to this day, the problem has not been solved. DO-178B wants you to solve it.

Our favorite question as software development managers is to ask a new guy coming out of school, "What is software design?" If the first thing they talk about is code—C++, programming—don't hire that person. That's not software design. It's programming, and software engineers confuse the two. Programming is learning how to *use* a hammer; while design is knowing *which* hammer to use. They know how software is put together, but not necessarily what it does. Software design represents how it actually interacts.

Requirements are easy to detect. Requirements versus design, what versus how. That's what we teach. It becomes blurred because the person who's writing microcode, the software language you're using, that's his requirement. So as you go from left to right in that waterfall the word "requirement" is blurred but, in general, we call a requirement "what" versus "how".

As the development process goes down the waterfall, you define pieces that 178B is looking for. You finally get to the source code level, doing coding and integration. FAA doesn't care what language is used (but we recommend ADA, C and C++). If you're working in C++, FAA provides a whole slew of guidance because C++ and ADA are dynamic. The runtime kernel in ADA has dynamics, and recall that DO-178B requires you to be deterministic. You can't be both.

Determinism is the opposite of dynamic allocation, dynamic stacks and dynamic heaps, which let the code and runtime kernel make decisions, interrupts or hold up the process because a printer somewhere is waiting. You deal with this very carefully. Even though coding is allowed in any language, be aware. We still like assembly language because you can't go wrong. The code has to comply with low-level requirements.

Out of this process you get source code and executable object code. There must be a review that says the source code complies with your design or high-level or low-level requirements as they are specified.

Conformance with standards means a review to make sure code conforms to your design and coding standards. There must be evidence—bread crumbs left behind—in a checklist of what you actually reviewed.

**Traceable low-level requirements**: Before code is generated, what do you trace? All that's required is that traceability exists at the end of the project, when you do the conformity review. However, the trace matrix is, in fact, a tool to do metric assurance that you are capable of tracing. Any software engineer using good development practices establishes complete traceability from the onset of the project through each phase. Trace all the way through because that trace is a management tool. Not only is it good for coverage assurance and metric management, it helps proper life cycle transitioning, a DO-178B requirement, from one phase to the next..

One transition point is going from design to code. You should have, at this stage, standards, reviews and a traceability review. Is the code accurate and consistent? These are objectives of the coding phase, found in Appendix A of 178B. Somewhere you specify how to determine accuracy and consistency. And, absolutely, you must have traceability during the code peer review because reviewers must use that traceability to ensure that the code implements the requirements it traces to, and only those requirements.

That's the standard development life cycle. One company represents its entire life cycle via a spiral model, and its representation is the basic waterfall, done again and again until it's right. Planning is left out because it was already done, but they do requirements, design and code. By its nature, DO-178B does not exclude such a software methodology; practitioners are allowed to be almost as inefficient as they care to be. But again, the artifacts need to be there comes the conformity review.

Every time you create a baseline, the DER looks for evidence that the defined process was followed. If you change code he expects to see reviews again of that code, of requirements and test cases—every single time. The 178B process wants it that way and is written so you have to show that evidence only once at the end. But it has to be present all along the way since QA, and a pro-active DER, will be auditing to ensure it.

The following table shows the typical gap between a "good" avionics development organization (performing at SEI CMMI Level 3) and that required by DO-178B, for each of the basic software life cycle activities. Note that quality assurance, configuration management and software development typically have smaller gaps, while unique DO-178B activities have a correspondingly greater gap.

| **Certification Activity** | **% "GAP"** |
|---|---|
| PSAC | 80% |
| QA Plan | 20 - 30% |
| CM Plan | 10 - 20% |
| Development Plan | 40 - 50% |
| Software Verification Plan | 60 - 70% |
| Safety Assessment | 80 - 90% |
| Requirements Definition | 20 – 30% |
| Design | 10 – 15% |
| Code | 5 - 10% |
| Functional Test | 5 – 10% |
| Structural Coverage Tests | 90 – 100% |
| CM | 10 – 30% |
| QA | 50% |
| Tool Qualification | 100% |
| Checklists | 30 – 50% |
| Reviews | 30 – 50% |
| Audits | 30 – 50% |
| DER Liaison | 100% |

# 23. Verification Revisited

Assume you have the code debugged in the lab and think it works. If you now write documentation to fit the code---this could be the quickest way for the DER to be sanctioned by the FAA, render the QA person worthless and prevent the company from getting its product certified. We don't want to write code, then the documents, do the traceability and assurance process, then present it to the FAA. That is what people have traditionally done more than once with TSO certifications because a TSO only calls for information at the end. The FAA caught on to this and doesn't like it anymore. In the precursor to DO-178B, DO-178A did not expressly prohibit this practice. But for over a decade DO-178B has been in effect on commercial avionics projects and it specifically cites the need for, and adherence to, a software life cycle model and transition criteria. DO-178B tells you that life cycle process is supposed to be followed as you develop. It shouldn't be neglected during development then resurrected at the back end after the code is written, although some companies have done that and had their products certified. That's not the objective and it doesn't contribute to quality or safety. And, frankly, it doesn't contribute to reusability, maintainability and cost-effectiveness downstream as most of DO-178B will when it is followed.

## Verification

Evaluating results of a process ensures correctness and consistency for inputs and standards. It's not just testing. Testing is one methodology of that verification. Most of the time verification means looking at correctness and consistency, mostly with reviews. One of our favorites to pick on is analysis, repeatable assessment, which goes something like; "I gave it to Joe to analyze." "Oh, did you analyze it?" "Yes, I analyzed it." "How? What process did you follow and where is it written down?" That should be part of your standards.

Remember that you have to specify your regression test strategy and an overview of that process within your verification plan. Do you perform regression testing? It's impossible not to. On the over 250 avionics projects we've contributed to, not one was perfect and unchanged throughout development; imperfections and changes begat regression testing. Was the regression testing you performed in accordance with what you specified in your verification plan? The best kind is to repeat it all. You can't go wrong because the FAA loves it, we love it---and management hates it because of cost. So most DERs, the FAA and reasonable people allow you to regression test only that software that was changed or could potentially have been affected by the changes. We all state that in the verification plan. How, then, do you determine what tests, requirements and code to regression test?

If an element of code changes, first plan on regression testing that element of code. Then use the traceability matrix (you're always keeping that updated, remember?) and follow all the trace paths for any element related to the items that changed (requirements, design, code, other tests, etc.). That's what your verification plan should state, albeit with more specifics. There are many types of analysis. Data analysis can be done so data associated with a piece of code in a module, wherever that data fans out, all code with that data, let's say global type data or pointers, needs to be regression tested. How do I know which requirements to repeat? If the traceability matrix is done, look at all code just analyzed, go back to the requirements, find test cases for those requirements and re-run them. Nothing more, nothing less. You know where to start and where to end. There's no ambiguity and it's repeatable.

In reading review minutes we may see comments like, "We analyzed it." "What'd you do for analysis?" "Well, let's go ask Jim." Well, that won't work.

**Review qualitative assessment of correctness.** A review is a qualitative method, an assessment of the individual examining and evaluating the item under review along with related documents. Reviews require engineering experience to determine whether the item being reviewed is correct. Don't look to QA to assist in reviews; their audits are not technical, remember? QA ensures your process is followed, not that your avionics product is technically correct. That's the dichotomy; you can't have it both ways. A QA person who performs process reviews along with technical reviews probably isn't performing at the required degree of rigor. So the FAA, via DO-178B, relegates (or "promotes") the QA reviews and audits to be process oriented; technical reviews are subservient. Review, however, doesn't only mean, "I read it. I did a desk check." Yes, I reviewed it and I've signed off on it to indicate I think it meets all criteria and is appropriate."

What did you review to? The FAA wants you to think about DO-178B and what you're reviewing to. What criteria does it have to meet? That's where checklists come in; you really need detailed checklists.

Test is not a qualitative process. It's a definitive process, whether recognized as such or not (and some smart people would disagree with this). Tests pass or fail. A test is repeatable and can be quantified. It exercises the system and satisfies specific requirements. It determines if something worked or not. It has expected inputs and outputs and expected processes. It can be done over and over again for the same answer. So once the coding and development process are done we've got the code.

What is code by the way? What is software? What is the source? It's a document, only written in a different language. All you have is the source and the source doesn't fly.

Integration testing is where you compile and check the integrated software via testing which includes the target hardware or a replication. We're trying to check the executable so test cases and procedures are written in order to run the tests and get results that replicate the real world. At that stage you've done another transition; from coding to progressing to the compile stage and removing errors. Hopefully, a static analysis tool such as Polyspace was used to remove runtime errors out of the compile stage and you're going to make sure test cases are correct. Then you do reviews to ensure the results are correct, as well.

During verification we do reviews and tests, with analysis of areas that we cannot discretely test. Unlike military testing standards, 178B never has demonstrations as a verification method. Incidentally, 178 doesn't talk about inspections. Software's not inspected, it's reviewed, although the words "inspect" and "review" are often used interchangeably. The appropriate term is "reviewed," even when doing an inspection. After all, you can't review without inspecting, but you can inspect without considering review criteria. So, in DO-178B you do reviews, not inspections, because your reviews are always to a detailed checklist.

Now wait...we just said you do reviews to a detailed checklist, but where do you obtain such checklists? They're in the back Appendices of DO-178B, right? No, those are not checklists; they are simply key aspects for each life cycle step that needs to be considered and referenced within your plans. Then where do you obtain checklists? Why doesn't DO-178B include checklists? Remember, DO-178B was developed by industry and the consensus was that no company would force its own culture upon others via DO-178B. And your checklists are an embodiment of your culture. Also, in the litigious USA, if the FAA provided checklists and you satisfied them verbatim, yet suffered a plane crash, whose fault is it? Some juries might be persuaded it was the fault of the checklist, hence the FAA, who supplied such checklists. The FAA is smarter than that on both counts; each company needs to develop or procure its own checklists which reflect its development practices and culture. Can you buy checklists from third parties? Of course, this is the 21st century and almost everything desirable can be procured for a price. We use HighRely's checklists, so are biased in that way, but other companies also sell checklists. Use your own due diligence to determine which source to use.

Would you ever use demonstration in a 178B process? No, because it does not meet DO-178B's requirement for determinism. What one persons observes via demonstration is not necessarily the same as another. Demonstration doesn't objectively prove that errors were detected, and you cannot repeat that demonstration and necessarily prove that you obtain the same results. There is no review, it's just a demonstration. Let's say a light blinks. We don't know why, but it does and perhaps the blink indicates a problem we should investigate. With demonstration, we don't know, so it's simply not used.

Military processes 498, 2167, 56, 55 defense and other IEEE-type standards allow testing via demonstration. If that's the process then turn that demonstration into a test. You had these inputs, ran this demonstration and obtained these outputs as you went through the test phase's waterfall life cycle. Do review stages at every transition point. Don't forget to tell how you determine consistency and accuracy. Have a statement that says how the checklist works.

You'll do reviews, analyses and tests at this stage, which are all objectives of 178B. How do you determine accuracy and consistency in that review?

The executable is also checked, with source-to-binary correlation required for Level A, but this is typically a test. The high-level to low-level requirements should be traceable within every baseline, and the baseline verified to ensure it meets those requirements. You'll execute all your hardware dependencies, I/O, BSP and RTOS on the target so you'll inherently provide compatibility with the target.

If it's executing on the target it's compatible with the target. If the test is for the executable not on the target, can you say it's compatible with the target? If you're doing tests and you're taking credit, and you do them on the simulation, it is allowed. One of the checks you shouldn't see is compatible with the target. You haven't checked that yet. All you've done so far is checked functionality, requirements to functionality.

You're reviewing the requirement that traces down to the design, and the code that executes it, and reviewing the test cases that functionally verify that requirement. Hopefully, you're using the same compiler and target binary for the verification so you obtain valid results. You can obtain credit for much of the structural coverage achieved during that testing by utilizing the target binary along with a structural coverage analysis tool during functional testing of those requirements. If I use a simulator for that testing, I'll want to repeat a subset of those tests, perhaps 5-10% comprising a representative sample, on the actual target to prove validity of my simulation environment. I'll show that this representative set of tests actually executes properly on the target, as well. Regarding simulation testing, most fault testing type handling is going to happen on a simulation because it's difficult to obtain fault insertion conditions and some robustness testing in a target environment.

Regarding robustness testing and hard-to-create error/boundary value tests, try to avoid having to test them on the target. Most target environments tend not to support invasive testing to determine if the software meets the requirements. Ideally, you would have a target testing environment which allowed for such fault insertion and timing control. However, that is very rare to achieve. You'll do them in a simulation test environment. Testing on a large system embodying 5,000+ requirements and 50K+ LOC typically has test cases split between a simulation environment and target environment. You might do 3,000 on the simulation and 2,000 on the target. Of those 3,000 on the simulation, take 500-600, re-execute them on the target and show equivalency; e.g. that they still perform the same on the target as they did on the simulation. If at Level A, you have to show that comparison anyway. If it's Level B you don't have to show that source-to-binary comparison, but every DER probably wants to see that done because we don't trust it when you run tests solely within a simulation. It's not what's going to be flying, but we still want to see it. It's acceptable to show that the majority of requirements have been verified on the simulation. We're logical and pragmatic people---we know that the simulation, to a degree, represents the real world if built properly. We know you're testing functionality. The only question is, does the target work the same way?

If the simulator is an instruction simulator, then we're not going to validate that instruction set simulator. The reason is, we show the same results on the target that represents that instruction set, since the instruction set simulator executes the same binary. If you are doing DO- 254 you can use a model sim; it simulates FPGA operation and how the CPLD would actually work. The model sim may require tool qualification in that respect. Model sims exist and we've been using them successfully for years.

How do I know it works correctly? We run tests on the target and show it works the same way, with the same results. Now the question is: what percentage of tests must be re-executed on the target to show acceptability of the simulation? A hard-core DER or true FAA representative would say, "Run 'em all." But they're reasonable people who know we are worried about cost and schedule, so you can run a representative sample. Now, we've all learned about requirements, but what is a "representative" sample? There is no right answer because it's a negotiation. Be sure to begin that negotiation up front and obtain agreement. If you wrote in your verification plan "10%" and the officials agreed, be sure to point out that agreement. You shouldn't be required to do more than 10%, although we know of egregious situations where it ran 30-40%+. .

You have a right to talk to the FAA and disagree. That's not advisable, but you have the right. It creates tension that's better avoided by negotiation; remember, you can win a battle but lose the war. And if your relationship with the FAA or DER ever reaches the status of "battle," you've lost anyway. Long-term success occurs via win-win negotiations, not continual confrontation.

Don't argue with the DER or the FAA. We've heard FAA people say to DER's, "We didn't approve your DER to argue with us, but to go out and do a job that represents us." Remember, your DERs, even if they're your employee, are representing the FAA whenever they have their DER hat on. But if you have a DER who doesn't ever "discuss" (negotiate) with the FAA, get rid of him because he's probably not representing your valid interests. Almost always, there is a little negotiation, and some sense of reasonableness. You want to challenge a few points (and just a few), but may not have to. On target test re-execution, just rerun a representative portion of the test on the target. How much? Define a logical representative sample within your verification plan, then negotiate to obtain agreement.

Test objectives verify that software meets its requirements and is done by pass-fail. Use multiple test cases and procedures traceable back to the requirement. Obviously, if it passes, it's correct. Identify with a degree of confidence that if errors exist, they have been identified.

## Commencement with Software Level

The test doesn't tell whether software, from a structural coverage perspective, executes correctly. All that structural coverage analysis tells is a metric of how well you wrote and verified the requirements with respect to completeness. So commence to the software level; did you write tests that were robust and covered the requirements? Did they test all equivalence classes or do all normal testing? Did the tests take all required (per the criticality level) paths?

Software requirements test generation covers hardware, software tests, software integration tests and low-level requirements tests. You should have all three types. They can be segregated into those situations but, realistically, we're looking to test interfaces, functionality requirements and derived requirements. What that means is to go through your trace. There should be a test for every type of requirement, whether it's a functional requirement, high-level requirement, low-level requirement or derived requirement. In some cases there are standards, which require tests to show you've met them.

Coverage analysis determines what software functionality is not tested. If you have not fully elucidated your requirements or fully tested them, you'll have an undue amount of code which isn't covered by tests You first perform a requirements coverage analysis just using your traceability matrix to ensure you've allocated all your requirements to tests and that the tests, in fact, cover them. This should have been done as part of the requirements process if you're smart and developed test cases at the time of requirements development. If you weren't that smart, or enjoy wasting a little extra money on your verification effort, then you wrote test cases too late to be of use during the requirements review so you reviewed those functional test cases later. Whichever method you chose, you should have reviewed the traceability matrix as part of the test case reviews and all those requirements should have been covered. Regardless, you'll want to re-review the traceability matrix to ensure it's fully up to date and you've covered all requirements with functional tests before initiating the structural coverage process. You're trying to get the "free lunch" and obtain structural coverage in conjunction with those functional tests.

.

In some cases there is a requirement to make sure the test exists, which is easily done with modern tools. Look for holes in the trace matrix and, if you find one, all requirements weren't covered. It's that simple. Do software structural coverage analysis and, hopefully, you instrumented all the code while running these tests. The results tell what code to look at if it's not covered, or to look at that same code because you didn't reach it but should have, and then perform an analysis.

Eventually, you're almost certain to find code which isn't covered. Now what? First, determine if the code should be executable, e.g. can it be reached during operations? This is a binary decision; no "maybe's" are allowed. It either can execute, in which case there is further work to do, or it cannot execute. If it cannot execute during operations, determine if it is dead or deactivated code. Dead code does not have a reason to be there so it must be removed...period. If there is a reason for that non-executable code to remain in your binary, then you have to remove it. If it's possible for the code to execute in flight, then you need to determine why it did not execute. If you can reach the code via testing, then reach it. Go back and add requirements that pertain to that code, such that the previously unexecuted code will be executed upon verification of those requirements. If it's virtually impossible to contrive a test scenario which executes this uncovered code, you may use analysis to verify it. If your DER is capable and on the ball, you'll need to justify such analysis to him. Regardless, document the analysis within the test results report.

Now you've found the uncovered code, added requirements, and there is no analysis which shows that it can be executed in flight, but test cases won't reach it. Now what? You have either deactivated code or dead code, and you need to decide which it is. Dead code is code which can not be executed in flight and which has no reason to be present; it must be removed. Why must it be removed since it'll never be executed? Well, can you truly guarantee it won't be executed? What if there is an undetected coding error, as simple as inputting the wrong character during the coding phase, for example, by errantly hitting the shift key. Impossible? Not at all; as every experienced programmer knows, shift happens. In such a case and others, it's possible for unexecutable code to be executed. It happens all the time with undetected pointer errors. Also, maintaining code containing unexecutable functionality increases the probability of future mistakes. So, dead code must be removed.

Now, you say that your dead code isn't really dead code since it has a reason to be there and you want to justify that reasoning. If you can justify it, and your DER approves (you'll need DER approval eventually, so better to obtain it early than later), then this dead code isn't termed dead after all, but is actually "deactivated". What is deactivated code? Code that is known to be non-executable during operations but for which a legitimate rationale exists for its inclusion. What comprises such legitimacy? Maintenance software, certain compiler inclusions, defensive coding, code meant for different aircraft configurations, etc. These are the most common sources of defensible deactivated code. Deactivated code must be reviewed, analyzed and defended within your test results, then it must be approved by your DER.

What if you use conditional compilation or compiler-supplied software libraries? Will these produce dead or deactivated code? This is so important it's worthy of further discussion. Conditional compilation does not itself produce deactivated code. The conditional compilation is used to selectively include, or exclude, source code for compilation. If the source code is not included for compilation, then it is not included in the binary. And if it is not in the binary, it cannot possibly be executed in the target, let alone during avionics operations.

Compiler-supplied software libraries may be different. Depending on the library, and your linker, the binary may include unused or unreferenced functions, hence it may include structures which cannot be executed. Such structures are deactivated structures and their handling is as described above for deactivated software.

Your coding standards should clearly state the rationale for removal of dead code and the potential rationale for inclusion of deactivated code. Software developers should be trained accordingly; it's easiest when they know they should never include dead code and minimize deactivated code.

If requirements and structural coverage are all you need, and DOORS is your requirements management tool, it's all in one spot. We never have to generate a document. It's a good idea, and testing is complete.

## Developing the Test

As an organization you've done the planning and are down to developing the test cases. How many people have test cases within test procedures, or test procedures within test cases? Depending on the company, that terminology is flipped. Some have one procedure, with ten or twenty cases and one parent procedure to execute it. Others say they have one case and repeat it ten or fifteen times in different procedures. It's a matter of terminology but be specific. DO-178B and DO-254 absolutely do not specify which terminology you should use. Just define your terminology and be specific.

You create a procedure, put in multiple cases, then initiate testing. Ideally, you developed functional test cases as part of the requirements review process and long before you had software. We keep harping on unit testing, but can you obtain credit for unit testing? We keep saying "no," but you could have developed a strategy to do some module testing, some integration testing, some ATP testing and some aircraft testing. All combined give sufficient coverage for all testing. When low level unit type testing is done on software modules (a fancy name for files) by the developer, we call that unit testing and such tests are informal. When low level testing of modules is performed independently and certification credit is legitimately sought, we call that module testing.

Module testing for certification credit is justifiable and sometimes desirable. Unit testing by developers as part of the informal code quality improvement process is also desirable; just don't try for credit of DO-178B test criteria via such unit testing. That's your strategy. You've got to have requirements covered and structural coverage analysis to prove it. One thing that's not acceptable is significant structural coverage at a low unit level because it's not done from a requirements or asynchronous system approach. The purpose of structural coverage analysis is to identify requirements that weren't written, or dead code in the execution. It's a metric.

Normal range tests are for input range, normal event interrupts and normal state transitions. When we talk about normal range tests, which include normal conditions and inputs, people often forget them. Typically, they think normal conditions and inputs just slow down the requirements. We talk about requirements all the time but the reason we get normal event interrupts is that they're a derived requirement. Those interrupts are generated by your design---a condition of the architecture.

There should be tests for all derived requirements. To make sure you have them, determine that the derived requirements are fed back into your requirements trace and into the requirements themselves; your requirements covered processes previously described will ensure you test them all..

**State transitions.** If a state diagram is part of the structure or a definition somewhere, expect a DER, as an engineer, to see tests that go from state to state to state. Exercise all such state transitions. Most likely you cover normal logic processing, and most of these state transitions almost every time, simply by writing functional requirements tests from detailed requirements. Just make sure you have complete coverage of all your possible state transitions.

We talk about doing high-level tests but not unit testing. One thing you want to test is that, in fact, something like a loop that has a counter goes through its initial to final value. What happens if the counter goes astray or if, in fact, the final value isn't the same as the value desired? You need to create some type of test to potentially get to that point. Do you need to check all code to make sure you know wherever there is a flawed loop? Only one test is needed; the boundary value test of that counter.

If there's an error and a pointer points to nowhere, what happens? There should be a test for that kind of error. A lot of people say they have built-in test that actually works---say, a bit that's constantly checking memory, or has a watchdog timer that checks flow. Should anything like that ever happen we're going to reset. It's a valid answer.

We're not going to check what happens if the For loop (that little counter) becomes corrupt and how the code is affected because everything goes wild and milliseconds later it resets. It's a valid answer but you need to test it and see the reset happen.. The reason these tests are forgotten is because they're not in your requirements. They typically are derived requirements that come out of the design. If they're in the matrix you're obviously going to write tests for them.

If you use a modern RTOS such as Integrity-178B, you'll want to incorporate the time-partitioning feature to prevent timing overruns. And you'll almost always want to use a watchdog timer capability to check for a hung microprocessor or even a process that exceeded its specified time allocation. Most BSPs support a watchdog timer whereby the processor maintains a counter within a register that is regularly decremented and monitored. Your BSP and application software should regularly strobe it, which means reset to a large, non-zero value. Then, if it should ever decrement to zero and cause a reset, the implication is that your processing is hung because the task assigned to strobe it to that higher value was unable to be executed due to a delay or hang up elsewhere in the processing stream. *Voila!* Successful watchdog timer operation.

**Exception handling.** Operating systems have exception handlers, but depending on which RTOS you use, you should not assume they work. If you don't have a COTS-certifiable RTOS, you may want to write your own exception handlers. If there is an error, like a comparison misses, we log it and come back. You may not have a requirement for that; it might have been a derived requirement. Make sure it actually happens that way and, if you wrote that kind of code, which you should find during structural coverage analysis, make sure the code executes properly. If it does, the requirement should also exist.

**Performance tests and analysis.** This can be amazing. Let's say there is a requirement that executes when it's allocated within the cycle. We do a foreground and background architecture that includes a polling loop. It's a 30-millisecond cycle and executed in 20 milliseconds. Ten milliseconds are left for background bit development. That's a common architecture for an embedded development product. The requirement is to execute in 20 milliseconds with a 30 millisecond frame time. We need to assess the best and worst case. This is part of our robustness testing. Remember, an informal definition of robustness testing is "breaking the software"; so go ahead and try to break it. Fully load the system with all inputs and outputs and make sure it always executes in 20 milliseconds.

What does it do if it overruns? People sometimes forget performance tests. DO-178B, in the software accomplishments summary, says you give it characteristics of the architecture. What does that mean? How fast does the code execute? How much memory? How much stack is used and explain how it was tested to show it should overflow. Was it empirically tested? Did you use analyzers? Did you use interrupt processing? How did you determine where the timing sequences are in the software?

Tests often don't get developed because they're not part of the requirements. There's no explicit requirement defining stress testing, which is a methodology to see if the product has acceptable reliability. It's not required by 178B. Stress testing is a different process, where you try to get everything to work many times harder than normal. If it's a 35 millisecond frame time we overload it with data for 35 milliseconds and run it over thousands of cycles. We're trying to stress it at the highest worst-case intensity and see if it actually overruns.

The idea is, if you have that architecture, where you have a 35 millisecond cycle, a foreground and background loop, it is going to overrun 35 milliseconds every now and then. We know it catches it once. We know it might catch it twice. What if you let this run for many cycles? Does it ever overrun to the point where it never catches up but crashes? That's the idea, and most architecture eventually overruns them to where they don't catch up.

We always set aside testing that is destructive. Clock the speeds up, go in and pull pins, cut traces. You can do it in emulation and not destructively. If you're at Level A and want a higher level of rigor there may be some destructive testing but it is not required. Most developers in a lab have fifteen to twenty cards by the time they get to the end of the project. And seven or eight LRUs that go through various stages, as done in development. You can probably afford to break one because what happens after you've certified the product? They go on a shelf never to be seen again. So do stress testing and break some things if you're so inclined; just remember it's not absolutely required if you can use analysis to accomplish the same thing. It's an interesting exercise to pull a pin on the CPU and see what happens. How does your software handle it? Does it drop the A-bomb?

We worked on a major bomber project and did that kind of defensive coding. We tested with a two-star general sitting in the simulator flying the bomb run. He received the order of battle and was doing a great job. We were in charge of the displays, which worked perfectly. A giant screen above showed the action. The general is flying down the track, pickling off the bomb, doing a great job; "OK, let's try it again." So he pickles off the bomb once more as we watch, shouting, "Oh, man! This is really working!" The general wanted to do it so realistically that we would see the mushroom cloud when, all of a sudden, a banner flies across the screen saying, **"It's Miller time!"** Ummm....the general didn't appreciate it. We're not sure which engineer was responsible for that but we have a reasonably good idea. You can have fun, but don't put it in your test cases.

## Test Case Levels

Remember what is defined as a unit? If you're using 498 methodology of CSCIs, CSCs and CSUs, you've defined units. Test the CSCI. Test the CSC. You're probably going to do requirements testing at a CSC level. So for every CSC or every CSCI, do that level of requirements testing, do that level of structural coverage analysis testing and take credit for it. Then you can make an argument for a piece of the larger puzzle. You don't have to build the whole system,

We like to do the structural coverage test and requirements testing right there. Take all the code, compile it, integrate it, instrument everything and run all tests at one time. But that's *nirvana* and it doesn't happen that way in real life very often. Most likely you will do much CSC-type integration, and pull together large cohesive functions. Test simulators are acceptable at this level. When you finally get to the whole system, you probably want to do it on the target. There are a lot of advantages if you keep with high-level testing.

## Test Software Functional Requirements

Most functional requirements should be tested as a system, giving you more coverage per test and more realistic verification. If you could take the entire system, put it together and instrument it, that is the ideal case which is sometimes possible if you've planned in advance. If there's enough execution speed it's fantastic. Simply by turning it on---you've heard this before and it's true,

## Disadvantages

Low level tests are harder to set up at a system level. The reality is that there usually is not enough memory and processing speed for all cases, particularly if you instrument the code. All tests cannot be set up on actual hardware and there may not be access inside the LRU. You don't always have tool sets to peek inside the CPU and see what's happening. To do full-up testing at a higher level takes more effort, but may be worth it. Once the process is set up when you do regression testing, it's there and easy to repeat, and will likely pay for itself over the product lifetime

**Test results require more analysis**. What went wrong? If you are at a higher level and obtain the wrong answer, where in the code did it happen? It's a tough analysis that requires tools to determine structural coverage.

One system we worked on had four DSPs in chain and two power PCs. We couldn't put it together as one system and do structural coverage analysis, so we did one processor at a time. Well, DSP's are unique animals. We couldn't do the whole DSP but only a particular function. That didn't work, either. We ended up doing one module at a time, running structural coverage analysis about a hundred times.

If you don't automate testing, you run into this worst-case situation. There are not yet many tools available to perform coverage analysis on the DSP and instrumenting the DSP code. We had to write our own code to make sure our own tool worked. Three months later we finally had a tool. Now we tested a hundred times for one DSP and got a piece of the answer every time. We had to write a tool that collected all pieces of the answers, integrate them so we knew what code was actually exercised and what tests exercised what code. That's the reality. It's great if you can actually do higher-level testing but the reality is that it's hard to set up and do the coverage. We are pleased that we, as senior avionics engineers and DERs, could convince the FAA we did it right.

If requirements are written at a very detailed level it's very specific and you know exactly what to test. It's nicely confined and bounded. If there are only four development systems and we're working at a high level, that means four people are doing tests. There are only four systems but if one is at a lower level they can all sit on their simulation systems, PCs or workstations and emulate and simulate and perform a lot of tests. That's the power of doing it at a lower level. You don't need to develop a $1M automated system test set-up because it's all done at a high level, that is, at a systems level. You can write some simulation to do injections and debugging.

## Strategy

Your verification test strategy should be defined in your verification plan. Where you'll perform each level of testing should be specified. There are reasons why it's done at different levels and you need to describe it. You perform low level testing because you're doing algorithm failure checks. It shouldn't be haphazard. In summary, the test phase under DO-178B is not trivial. About 40-50% of the development budget is spent testing, verifying and analyzing. If it can be brought down to 25%, that's world class.

## Analysis

When test contents are confusing there may be several meanings. When analysis is used, don't just say you've analyzed it, but tell what it means, what you expected, and what you found. It must be repeatable. If your partner uses the exact same analysis--- she should obtain the same results.

There are many strategies to testing. Decisions are based on the type of development. If it's an operating system, we do a lot of low-level testing. If it's only an interface or board support package, we'll do lot of low-level testing. If it's writing a flight management system you want a lot of high-level testing.

There isn't one right way to do this. A key thing about verification; if you believe what you've done is correct and it deterministically and consistently meets the intent of 178B, then stand by that belief and make your argument. Write it down. Don't let somebody baffle you into doing something that is unnecessary.

# 24. Project Organization & CMMI

Why does it take so long to develop life cycle process and data expected by DO-178B—with everybody reading the document to find a shortcut? There are four types of people needed to do the project; those who develop it, those who verify to make sure the developers are doing it correctly and validate it for correctness, then QA people who ask if the standards were followed, and everything is defined, and finally, the DER. That way, we can't confuse each other.

> **Project Organization**
>
> What is the minimum number of persons needed on a DO-178B project?
>
> What is the optimum number?
>
> What does the organization look like?
>
> Does CM require a separate person?
>
> S228/182

Does CM require a separate person? We recommend that if you have a large project, e.g. more than thirty engineers, you have one, because CM is a discipline unto itself. The tools are complicated today. You will need manual processes to assure that what goes in the CM system belongs there, and there are no unwarranted changes, in addition to using the "CM Person" for the build of the software.

There are four major facets to configuration management: problem reporting, change control, statusing and version identification. Configuration management, especially on large programs with more than thirty people, should designate a configuration management person. Even with a five- to seven-man team developing a piece of software, how many times will you rebuild and recompile the system? You should have the CM person doing the rebuild and recompile so it happens the same way every time. It will then be known if somebody did a checkout at midnight and made changes that cause conflict. It's an extra person but he might be worth it.

What does a typical mid- to large-size project team look like? The diagram below is fairly typical. But remember, DO-178B doesn't define project teams, budgets or schedules. You define them.

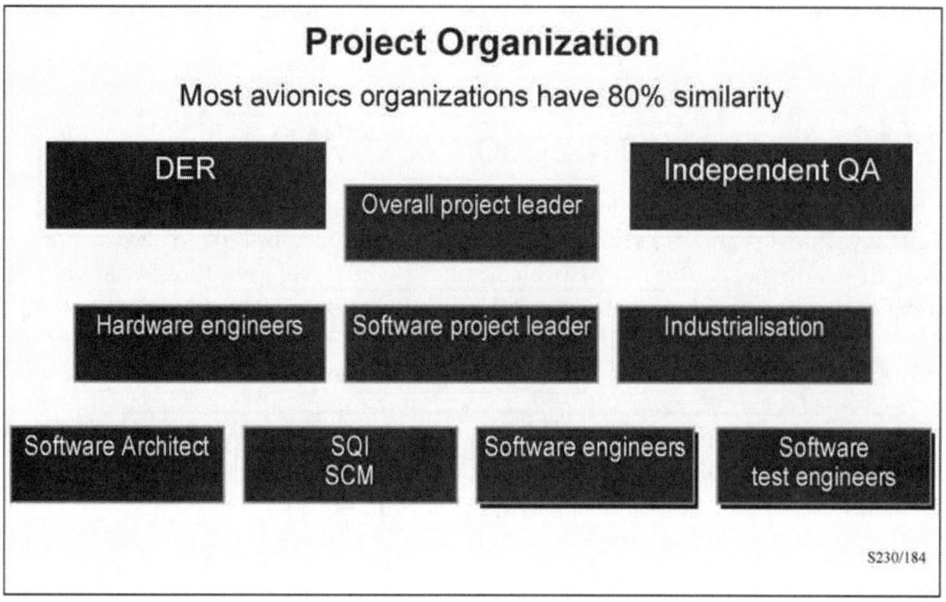


If you're not familiar with CMM, it was developed via Carnegie Mellon University, which developed the Software Engineering Institute (SEI). The man in charge, Watts Humphrey, was asked back in the 1980's by the Defense Department to figure out how to assess vendors on large software development projects. The DoD was letting contracts that didn't say the system had to work, only that there were certain requirements. They needed to figure out how to get capable companies who would build something that would work.

**Carnegie Mellon University**
**Software Engineering Institute (SEI)**

Devise a method of assessing and grading software organizations

"Capability Maturity Model" (CMM; though not coincidentally similar to the university's initials)

Why that acronym?

S232/186

Watts, at Carnegie Mellon, was asked to develop a method of assessing and grading software organizations and the result was the Capability Maturity Model. (The acronym—CMM—is an implicit acknowledgement of its affiliated institution, e.g. the Carnegie Mellon Model).

## SEI's Findings

Software is a creative art but process-driven

Good organizations produce better software

Software organizations have many objective criteria

Organizational maturity can be assessed and ranked

S233/187

The institute found that software is a creative art but process-driven. So we software engineers create art, but put in a process to control and predict that art.

Next is the question of what is a good organization? The CMM model tries to describe it in quantitative terms. Some qualitative reviewing and assessment are done, but it tries to define a good organization. There's a correlation between that description and its maturity, and the quality of software that comes out. What the Defense Department means by quality is that it works, it's on budget and on schedule. Is that what 178B is trying to achieve, as well, and the FAA trying to achieve?

## SEI Devised the CMM

Capability Maturity Model: CMM

Ranks software organizations 1-5 (5 = Highest)

Evolutionary: Improves one level at a time

Impossible to skip a level as you grow

S234/188

The rankings they came up with for the CMM model are 1 to 5; 5 being the highest, 1 the lowest. How do you become a "1"? If you say you write software you're a 1!

Typically, an organization begins as a 1. They're *ad hoc* and go to Level 2 if they generally have plans in place which personnel can and should follow. Then they do the assessment and may achieve Level 2. A little more is learned and their processes are more repeatable. At Level 3, the company is managed and consistently repeatable, then goes on to a 4 and 5 which include continual process improvement and metrics

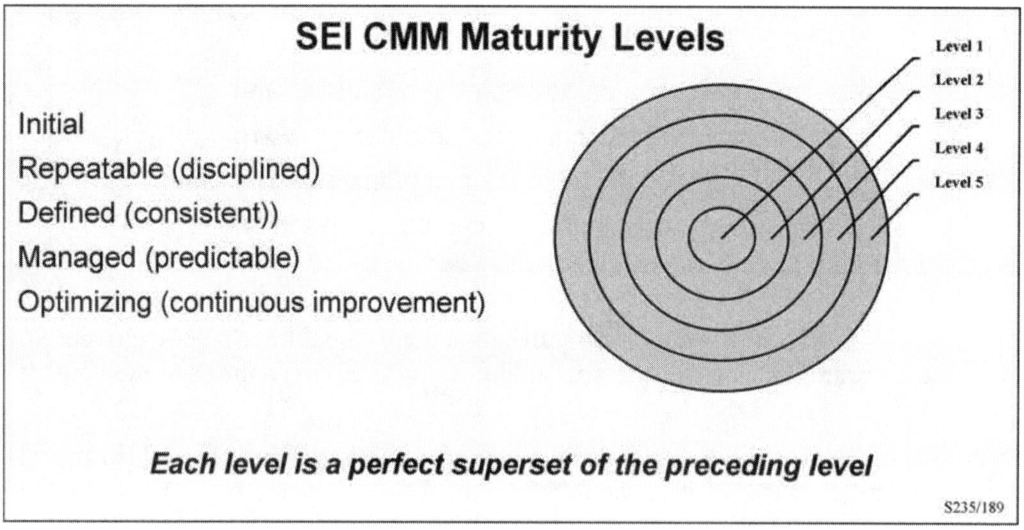


Level 3 is consistent. Not only are you repeatable and have discipline, but are consistently the same and managed. Level 4 is predictable.

**Optimizing (continuous improvement)**. If you read that in the CM model it means you're working together to make it better. In that system be sure everybody's following the process. If you are all improving and optimizing, you need someone else to check. At Level 5 you should have a QA organization. At Level 1 the lowest processes are defined and they do not require independence.

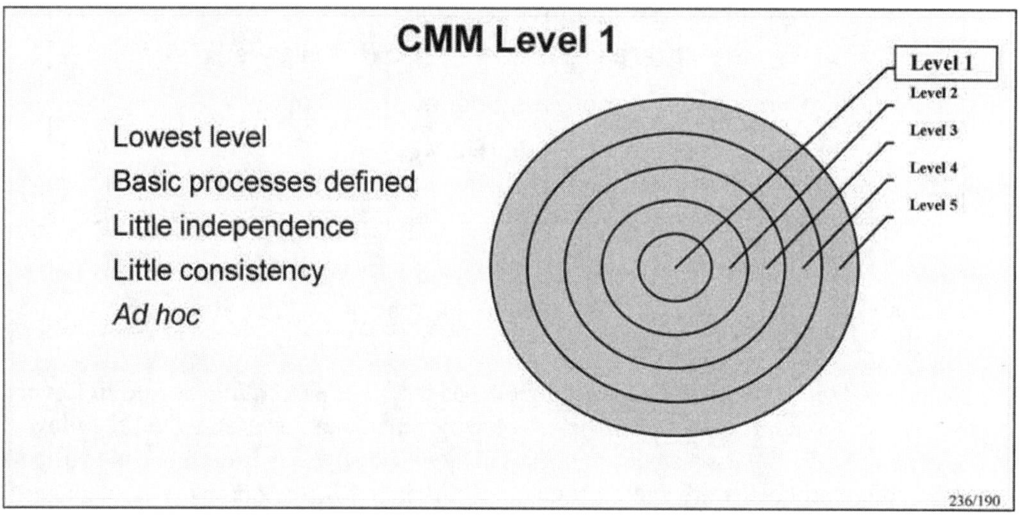

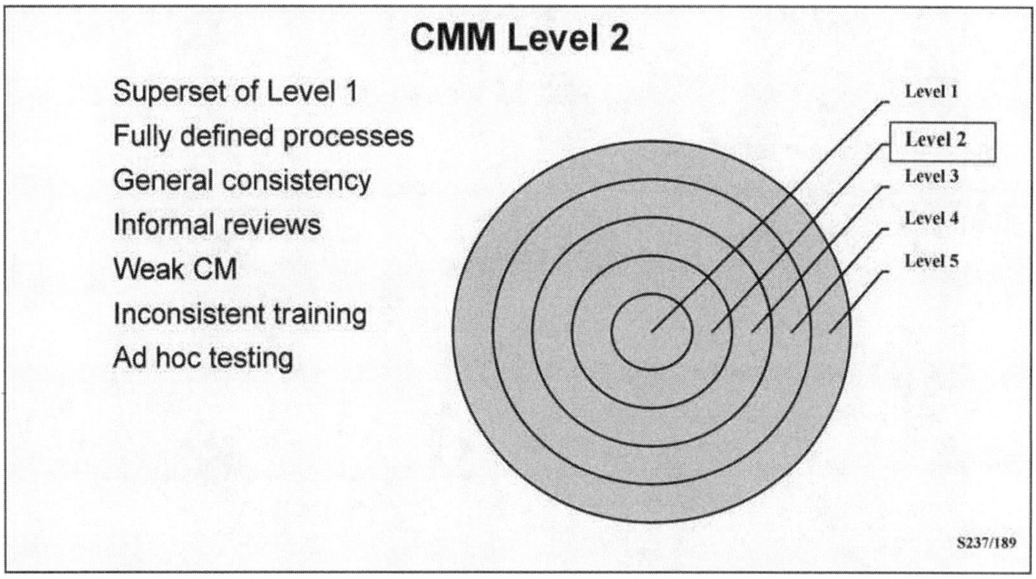


Level 1 is usually a commercial development organization where the person writing code is also doing verification of that code and the CM (and maybe sweeping out the closet). There's little consistency, going from one program to the other program. And there are many weaknesses but you don't even know where they all are. Very *ad hoc*.

Level 2 has a software development process fully defined by ISO. You don't *have* to follow it but ISO defined it and somewhat applied on the project.

General consistency is going from program to program, being sure everyone talks the same language and using the same procedures and processes. There are informal reviews, informal analysis and methodology. Someone says, "We had a review of the document and, yes, we all were there—thank you very much. We think it's good." That's a weak process but actually fairly typical for less mature companies developing software. Typically, we have a rudimentary CM system and some modestly defined processes and checklists. The developers check code in, check it out, make changes, etc. Nobody checks if the changes are appropriate or not. Come back a year later and go through the VSS (Visual Source Safe) system and you can't find half the data. Or you may not have the right versions. Maybe no build scripts. The executable may not match the source when you compile it. There may be some CM and QA but they are not consistent.

**Inconsistent training (if there's any at all).** Good training is important. *Ad hoc* testing, usually done as a beta, is where you do some debugging then build a system and hand it over to someone to whack on it, pound it or do some basic user-centric testing. Much of the software you run on your PC was developed this way. But it's still *ad hoc* and not really controlled.

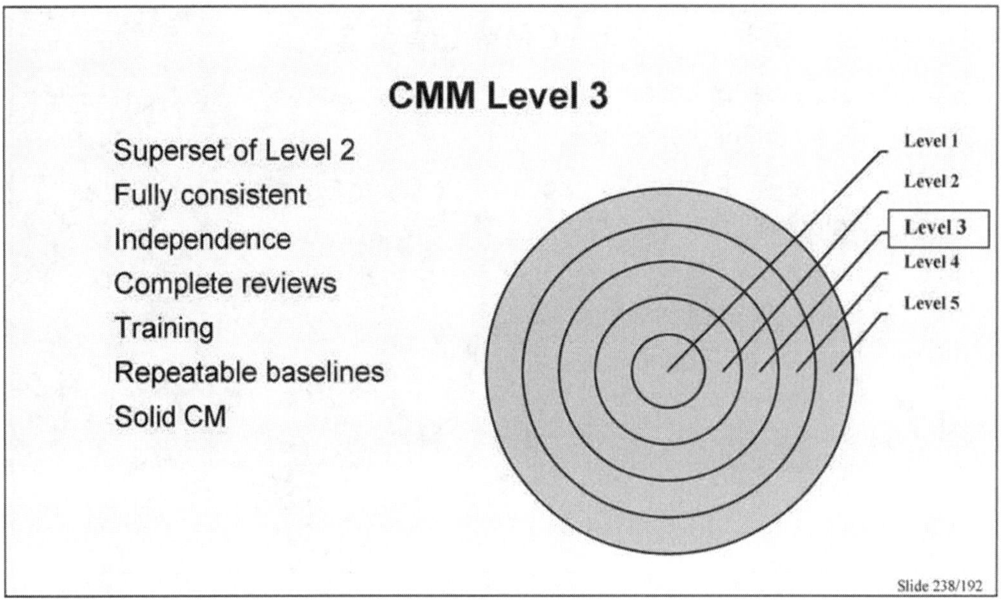


Level 3 (after you've progressed from Level 1 through Level 2) is fully consistent. When you go from program to program using the same processes and review techniques, the methodologies are consistent. Independence is fully defined. Testers consistently test all the code and QA is in place. Reviews occur for most major artifacts and consistent standards and checklists are applied. At this point, you don't want the developer to be the reviewer. It takes two pairs of eyes to make sure something is correct.

Complete reviews involving checklists are present at Level 3. A Level 3 company has training throughout for developers, management and verifiers, as well as QA.

**Repeatable baselines—good CM.** If you come back a year later you can repeat and build the exact same code. You can find all documents that go with the code and development cycle. CM is solid. You can trace it back and find all the information.

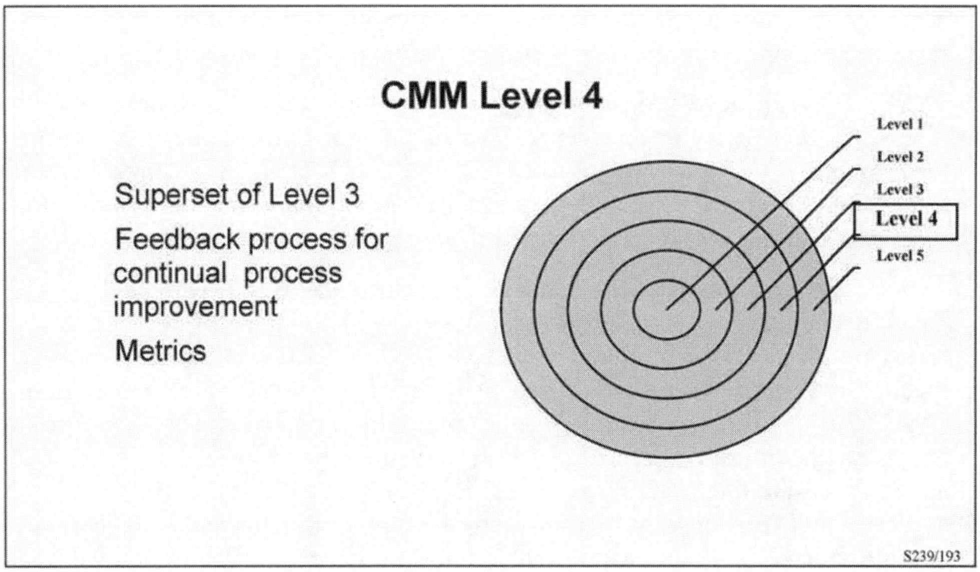


Level 4 adds the feedback process, continual process improvement and metrics. You start checking management, traceability, quality of the software, how efficient is the schedule, how predictive you are, how fast you can do burn-down rates and burn-down charts, and determine how many requirements take how much time.

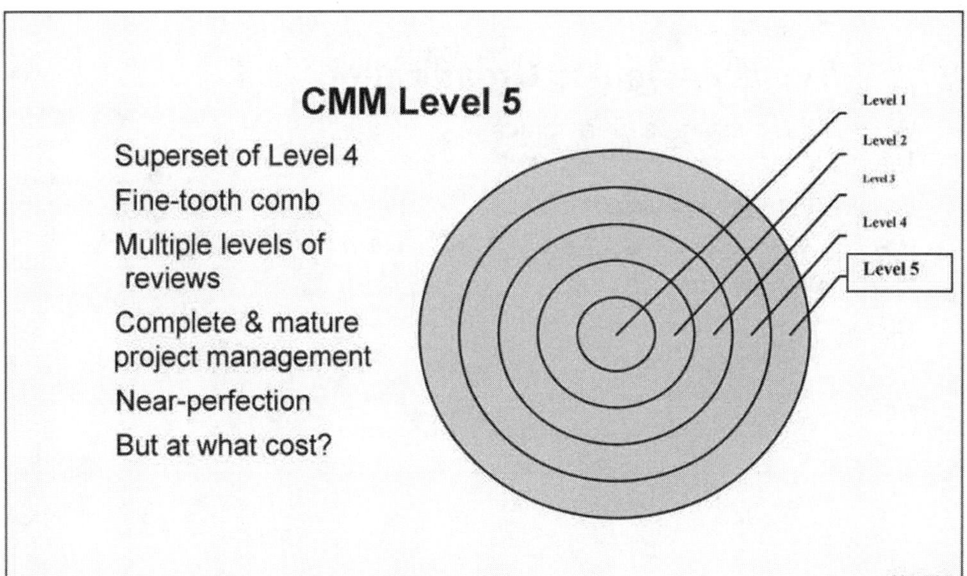

Level 5 (we've arrived at *nirvana*!) is a fine-tooth comb; everything's nearly perfect from a process viewpoint, with multiple levels of review at every stage and transition criteria. Complete and mature project management.

They all have project management training and know how to do all cases. It's near-perfection and expensive to do. Some companies say they're Level 5 but it may only be marketing spin. If a section in a company gets assessed as Level 5, it's typically a synergistic group that's worked together for years and have a good process going for themselves.

That is a basic CMM explanation. But what is a CMM explanation doing in a book on avionics software? Does DO-178B require, or even allude to, CMM or now, CMMI? Remember, DO-178B and DO-254 do not prescribe specific development methodologies and there is absolutely no mention of CMM/CMMI. But there actually is a correlation; avionics developers need processes, they must be defined and followed, and key persons (QA, DERs and the FAA) ensure those processes are sufficient with the result that they improve over time. So in answer to the question "Which level should an avionics developer be at?" the answer, in our opinion, is Level 3; clearly defined and consistently followed processes. Why not Level 4 or 5? While certainly acceptable, your processes do not need to continually improve because they must be good enough to meet DO-178B/254 in the first place, from the onset! After all, if you must continually improve your processes, what about the development occurring prior to such improvements; must it be re-done?

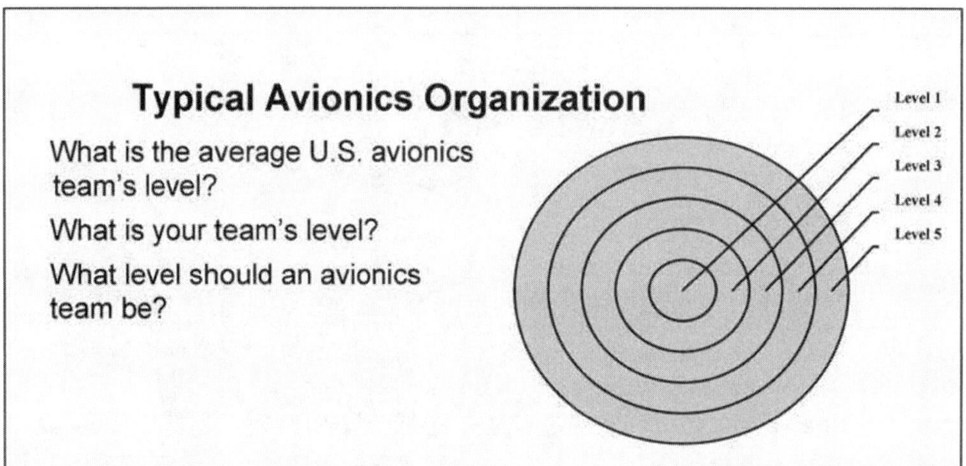

What level should an avionics team be at—Level 3? Remember that solid CM, independence and reviews start sounding like DO-178B. If you're a solid Level 3 CM organization, you probably meet 70-80% of what 178B is looking for. However, the average avionics team, in our opinion, is at Level 2. The authors of this book have worked with the world's top 100 avionics companies and we stand by this opinion. Is Level 2 a bad thing? Absolutely not; most airplanes today are flying perfectly fine software developed by Level 2 organizations. Then what is the implication? Truth be told, Level 2 organizations are simply inefficient and almost certain to fail meeting budgets and schedules. However, budget and schedule management isn't typically a part of a Level 2 organization anyway, so they don't even realize how inefficient they are!

Level 2 organizations simply require more horsepower to achieve the end result. Like a car with its brakes permanently engaged, a larger engine can overcome the added braking effect. But that engine is more expensive and uses more fuel. Level 2 companies require more energy re-doing their work and iterating along the path to certification.

Chances are you have competitors, and long-term marketplace success is based on competitiveness. Level 3 is where you want to be, with some Level 4 aspects thrown in (proof of training, quality metrics, etc).

# 25. PSAC
## Plan for Software Aspects of Certification

Let's revisit the PSAC because it's such an important document. While it is still a plan, it is really the embodiment of your certification strategy and process. You don't want to include too much detail which would only draw attention away from your key certification topics. The PSAC is your single most important software certification document and will certainly be reviewed by your DER and the FAA.

Major modifications to the PSAC are resubmitted for approval, but if the change is minor, it doesn't require a complete document re-do. That's why you have an accomplishment summary, where deviations are documented. The key is to keep the DER and FAA informed of even minor changes and let them suggest something, such as; "That change is okay, you can document it in the accomplishment summary." But if you change your approach, say, on major issues such as software tool qualification, redundancy, criticality level, etc, a re-release of the PSAC may be required, though minor deviations are not significant. For major ones you definitely want to revise the PSAC. If you thought you were Level C and the FAA or Safety Analysis calls it Level B, you need to revise the PSAC to reflect that, and all the ancillary changes which would ensue.

## PSAC Goals

The minimum software life-cycle data that must be submitted to the certification authority:

Plan for Software Aspects of Certification (PSAC, a crucial document)
Software Configuration Index (SCI)
Software Accomplishment Summary (SAS)

PSAC is the primary means to determine whether the software life cycle proposed is commensurate with the rigor required for the level of software being developed.

PSAC communicates completeness and consistency with the means of compliance agreed upon to satisfy the certification basis.

S227/274

The minimum life cycle data that must be submitted to the Aircraft Certification Office are the three documents mentioned above. The PSAC is normally approved by the FAA. Some FAA offices delegate approval to DERs, as well, but they want to see the three documents as a minimum. That doesn't mean they can't ask for anything else, or not have the right to come to your facility and audit every artifact----and they do that, depending on criticality of the system, novelty of the project, special new tools and processes you're deploying and fourth, and most important, your level of expertise in the DO-178B process. How many systems have you certified as an applicant? If you're new to this process and criticality level is high, they pay a little more attention to you.

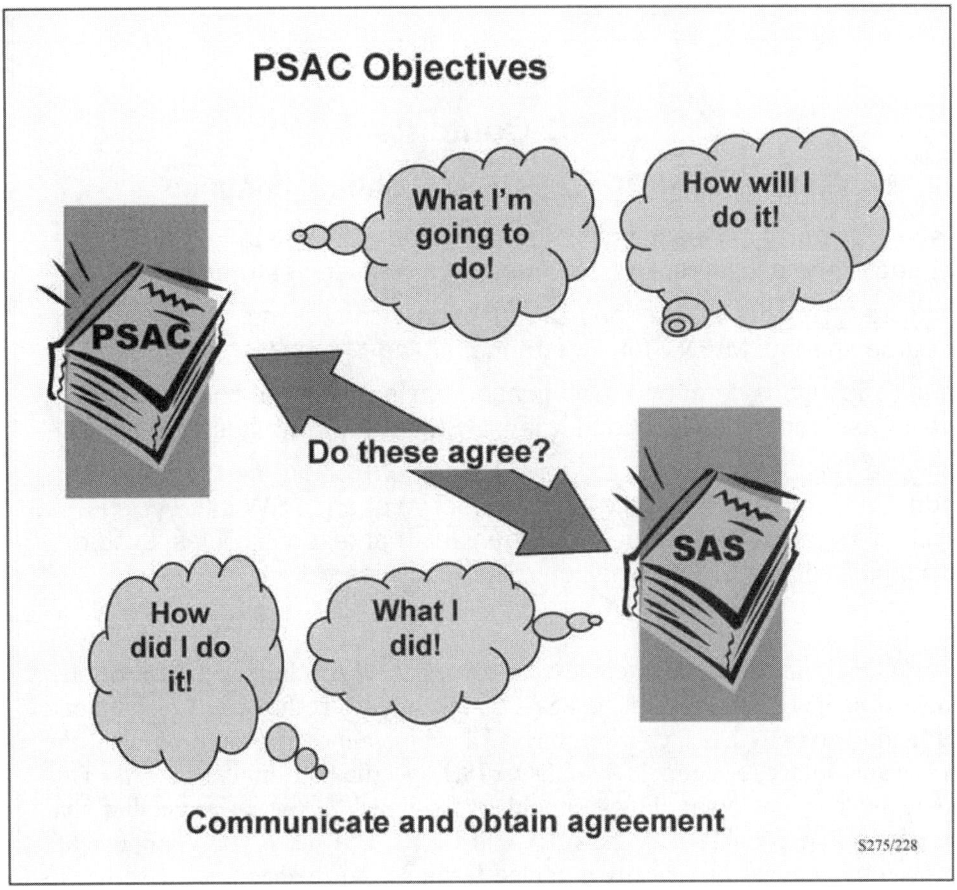


The PSAC, therefore, is the primary means to find out whether life cycle data and processes comply with DO-178B for the level you're proposing. That is the contractual agreement between you and the FAA or DER, and it establishes a working baseline. Try to get this done as early as possible to open a line of communication. You typically start with proposing, say, an STC program (supplemental type certificate) and apply to have the FAA generate a project number.

Next, you may create a program or project-specific certification plan. This is not related to software or hardware, but to the whole application. It defines your overall certification approach, TSOs you're going to need and FARs (Federal Aviation Regulations) which apply. It also points you to the PSAC and, if necessary, the PHAC (Plan for Hardware Aspects of Certification). Then you establish a PSAC for software and get a minimum working relation in regards to software compliancy. The PSAC communicates completeness and consistency with the agreed-upon means of compliance for satisfying the certification basis. This is a contract for complying with DO-178B for software and the approach you intend to take.

## PSAC Contents

### Section 11.1 DO178B; PSAC should contain:

System Overview – Describes system functions; allocation to HW/SW, architecture, processors, HW/SW interfaces & safety features

Software Overview – Describes SW functions; partitioning, redundancy, resource sharing, MVDS, fault tolerance, timing & scheduling

Certification Consideration – Certification basis, means of compliance, level of assurance, justification for level (SSA), SW contribution to failure

Software Life Cycle – Summary each life cycle (Requirements, Design, Code, Verification, SQA, SCM), how objective of each SW Life-cycle is satisfied, organizational structure, organizational responsibilities, system responsibilities, certification liaison responsibilities

S276

One question always comes up; Does the FAA have a recommended format or table of contents for the PSAC? The answer is they can't recommend that, and everybody has their own, but it's become almost standard on what the FAA wants to see; Section 11.1 of DO-178B specifies the minimum set. This should be met but other things should be included. Always remember that everything is negotiable with the FAA and DER. But when you communicate consistently and in a historically accepted fashion, you're ahead of the game, so why start the game behind?

Once the configuration index is done and you finished the program, put together an accomplishment summary. This states what will be done during the next X weeks, X months and X years in the software life cycle. The accomplishment summary says what actually was done, and it complies with the template format asked for by the PSAC

Some companies take their PSAC, call it an SAS, and turn the "will do's" into "have done's." That's not necessarily true because they didn't do everything, and probably made at least minor deviations. The accomplishment summary, however, must have special additional data about open problem reports against the project at the moment you're certifying. You may point to the configuration index for their content but you must have them somewhere, for example; what did you fix during development? What features did you add? Did you accomplish everything? Don't tell it in great detail. The accomplishment summary will contain them.

Section 11.1 of DO-178B states, as a minimum, that you need an overview describing system functionality, architecture, processors, interfaces between hardware and software and safety features you've built in. The software overview also has to be included. If there is fault tolerance, a partitioning scheme or redundancy, you may have to implement a dissimilar architecture. What is "dissimilar?" Back in the 1990's we worked on a program for a flight control system. The FAA said it was critical Level 1 software (DO-178A was in effect then) and there were three levels; 1, 2 and 3. We negotiated and ended up discussing an approach to keep the system critical, with the software at Level 2 or lower criticality.

How did we do it through negotiation? First, it was a flight control system and we ended with three redundant, identical LRU's on the aircraft that perform the same operation. The FAA said, "Well, that's not enough. Even though you have three boxes, they all have the same software. They could all fail the same way. What else are you going to do?"

We replied; "In each box there is a control that commands, and a monitor that monitors activities of the box and control functions. So it looks for data coming out of the control side and checks all hardware and functions."

"Not enough," FAA replied. "Again, there could be identical failures in all three boxes."

We said, "How about this? We write the control in ADA language and write the monitor in C." They performed the same functions and compared internally before comparing with other boxes. The FAA said, "No. We still don't feel comfortable about the possibility of failure, even though you're writing one in one language and the other in a different language" Then they said, "We require one more thing to give you approval to be Level 2 or lower criticality for the software, even though the system is critical."

What do you think they asked? Again there were identical boxes. Different CPU's already in place. Different compiler in place. Those were part of the negotiations. That's the similarity. Then they asked us to do the same functions written in reverse logic. So we had to define not only redundancy on how to reduce the criticality level by applying reverse logic on a dual channel-triple redundant system, but dissimilar architecture and reverse logic on the monitor channel. They went overboard and it caused so many problems they never asked for reverse logic again. (And this airplane is flying today, quite successfully. The avionics manufacturer, however, lost a fair amount of money on the development.)

## Section 11.1 DO178B
## PSAC Should Also Contain:

Software Life Cycle Data – Data produced and controlled, relationship of data, identify data to be submitted, form of the data and means of submittal

Schedule – Means by which visibility is provided to authority to allow for planning

Additional Considerations – Features that may affect the certification process:

- Alternative means of compliance
- Tool qualification
- Previously developed software
- Option-selectable software
- User modifiable software
- COTS software
- Field-loadable software
- MVDS (Multiple Version Dissimilar Software)
- Product service history

S277

Consider the certification level and justification for system safety analysis to justify the criticality level or design assurance level. Life cycle data summary for life cycle activities should include who will do the certification liaison work. A top-level schedule is also needed. What data will you produce and the relationship of these data to each other. You may not have previously developed software but do include it as a sub-paragraph to your additional consideration section of the PSAC and say that no previously developed software will be utilized. That way you considered it and the question is gone. Even if you are not utilizing any of the above, include a paragraph placeholder for each such aspect and simply state you are not utilizing it. You will make the reader more comfortable that you are aware of key issues and not merely hiding something.

What about the other plans, architectural features, detailed schedules, detailed tool and component descriptions? You have given each of these full consideration and made valid, informed and conservative decisions. So you want to include such details within the PSAC, correct? Absolutely not. The PSAC is not the dustbin for all your project details. It is a certification-specific document that references those other details within their respective plans and artifacts. Keep your PSAC lean and clean. How lean? 25–40 pages almost always suffice. Less than 15 and you can expect lengthy questioning downstream.

## PSAC Pitfalls: How To Avoid Them

Proceeding with development without agreement to the PSAC (could be very $$$ costly)

*Develop and submit early, get written agreement, early involvement of FAA/DER*

Lack of addressing all contents

*Use checklists, be complete and clear, get FAA/DER involvement early*

Incomplete or ambiguous

*Use checklists, be exact, get FAA/DER involvement early*

Modification of plans without update/notification

*PSAC is controlled and submitted, living document*

PSAC doesn't match SAS

*PSAC is a living document, modifications maintained and submitted, get FAA/DER involvement*

Design-level assurance not justified in PSAC (too high, too low---could cost you $$$)

*FHA, SSA, PSSA and system safety process is complete prior to PSAC agreement*

**Complete and submit PSAC at the *beginning* not end.**
**Avoid pitfalls by communicating *early and often.***

S278/279

Design level assurance is not justified in the PSAC. Don't try to put your FHA system safety analysis and all documents in the PSAC. These were done in an earlier system development process and system safety assessment. That's how you came up with an architecture.

The PSAC is not submitted until there is agreement on criticality level, system safety analysis and how we're going to be software Level 2. Then the PSAC is submitted. Try to get that done as early as possible and come up with the level. The PSAC shouldn't contain all the artifacts or details, but reference a set of data items and plans that deal with system architecture and software level, CM, QA, development and verification. Complete and submit your PSAC at the beginning. Avoid pitfalls of a plan without updates or notification. Companies go through development with an agreed-upon PSAC but make subsequent changes which may put them at risk. The PSAC does not have to match the SAS for every detail; it is expected that several minor deviations will be made during project development. The PSAC is a living document but there are areas which the SAS may require justification for why the PSAC wasn't followed. What are the open problem reports at the time of certification? They are referenced in the accomplishment summary. Again, get your FAA and DER involved as early as possible for understanding these items.

Let's talk about pitfalls with the PSAC and how to avoid them. It's important because once you get it in front of the FAA it's the first impression of how well you're versed in this process—and if you ask the FAA for comments, well, that doesn't make you very well-versed. It's like sending in your tax return, just signing it at the bottom, putting all your receipts in a box and saying, "IRS, you figure out how much I owe you." That's not going to work, or at least not in your favor.

You want to have done it all, show your competence, that no mistakes were made, and get agreement as early as possible. Don't leave things behind; have a checklist of items contained within the PSAC and, as a minimum, include all those items. Get the FAA and DER involved, since they've seen a hundred of these before and know what to look for.

Avoid any assumptions. Complete the PSAC and show exactly what you're going to do. Spell out some activities as described here. If you're planning on performing tool qualification, write down the name and purpose of the tool. Don't merely write in the tool qual section of the PSAC, "We will be using some tools and will provide the qualification at a later date." What are those tools and what is the date? Put in, "We plan on verification tools only. No development tools will be used. The verification tool is Vector Software's VectorCast for structural coverage analysis." Name the tools, but don't put in the version number. "Our approach for tool qualification will be…" then add just three or four bullets, citing the tools to be qualified (which should not be many) and that is all you need. The rest will be detailed in other documents later. That's acceptable so long as you have a solid approach and explain it succinctly. The same goes for each of the other PSAC sections and primary concerns mentioned above.

# 26. Tool Qualification

The subject of tool qualification is often misunderstood, although it is much the same for DO-178 and DO-254. The major dissimilarity between 254 and 178B is in flow charts, which will be addressed separately. There are two types of tools which may require qualification; development and verification tools. Verification tools are simpler to qualify because they cannot inject errors into your product. They can only fail to detect them. Development tools can actually inject errors if not developed or utilized properly.

For the purpose of qualification, tools are accessories to automate, replace or augment the manual activities cited in DO-178B and DO-254. Common verification tools are used in the correctness process and include test case generation tools, test execution environments, test pass/fail determination tools, simulation/stimulation tools, bus interface/capture tools, etc. Common development tools include code generators, modeling tools, compilers, etc. There are really only two initial questions you need to ask: (1) Is an item a verification tool or a development tool? and (2) Does that tool need to be qualified?

So first determine if an item automates, replaces or augments the manual activities cited in DO-178B and DO-254. All items mentioned above meet the definition of "tool" and should be cited accordingly. However, the real question is, do you need to qualify it? One of the most frequent mistakes we encounter is based on these simple questions. We continually see failure to qualify tools which must be qualified, and qualification of tools which is not necessary.

To determine if a tool needs to be qualified, there are three follow-on questions. They are one, two and three, and this is an "and" function. You have to obtain yes, yes and yes on the tool to be required to qualify it. Then you qualify the tool under a development or verification scheme. If the answer to any one question is "No," then you do not need to qualify it.

The first question, "Is it possible for the tool to insert a defect into the product, or fail to identify a defect?" If the answer is "Yes," proceed to the second question, where you usually get a "No." This second question is where you often eliminate the need for tool qual of many tools, including compilers.

## Software Tool Qualification

**There are two types of tools:**

**Verification Tools**: Tools that cannot introduce errors, but may fail to detect them.

**Development Tools:** Tools whose output is part of airborne software and thus can introduce errors.

**Do you need to qualify the tool? Pose these questions:**

1. Is it possible for a tool to insert a software defect or allow such a defect to go undetected?
2. Do you intend to skip verification of the tool's output as cited in DO-178B Section 6?
3. Are DO-178B processes eliminated, diminished or automated?

**If all answers are "Yes," you must qualify your tool as a test or development tool.**

S281

Let's look at verification, as defined in section 6 of DO-178B, using a compiler. Can a compiler insert an error into the code? Yes, certainly; a defective compiler would most likely insert an error into the code so the answer to the first question is "Yes." However, look at the second question: Do I intend not to verify my object code? I run tests against the object code, therefore, the answer is "No." I do not intend to skip verification of this step, therefore, the answer to this question is "No" and I do not have to qualify the compiler.

Now let's look at configuration management (CM) tools. They enable us to keep revision control, but what happens when checking the material into CM? Can a defect in the tool cause a defect in the source code? Yes, but per DO-178B, somebody is required to review the code and one of the first things to look at is the version number, change history and the artifact deltas, which verifies the CM. Therefore, as in the case of the compiler, CM tools also do not need to be qualified.

The major tools we normally verify fall into the verification tool category, specifically structural coverage analysis tools. There are test tools such as signal input generators and structural coverage analysis tools which alter the code; one instruments, the other analyzes results based on ports that were covered. The instrumentor is part of the verification tool. So you qualify the tool you purchased for instrumenting and analyzing the structure of the code for coverage. That's a verification tool, though. It doesn't inject errors into the end product because we don't fly with instrumentation so it's not a development tool for that reason.

| **Artifacts: For Development Tool or Test Tool?** | |
|---|---|
| **Development Tool** | **Verification Tool** |
| System PSAC | System PSAC |
| Tool qualification plan | Tool operational requirements |
| Tool operational requirements | System Software Accomplishment Summary (SAS) |
| Tool accomplishment summary | |
| System Software Accomplishment Summary (SAS) | |

S282

What must be delivered as part of the support data for tool qual? You already have a PSAC for the system and software. Now you develop yet another qualification plan for the development tool, say a code generator, that you do not intend to review against the model, or verify the code with the same rigor as for section 6 of DO-178B because you're using a qualified code generator. Then you will need a qualification plan with operational requirements for the tool itself. This has the same level of rigidity for requirement, design and usage as for the product that will fly in an airplane.

**Accomplishment summary for the tool itself.** Within the tool operational requirements you also include processes you utilized to develop the tool. Did you have and follow proper CM practices for the tool? Did Quality Assurance get involved? Did you verify the tool's functionality to the same level as you would have for custom-developed source code? Did you perform structural coverage analysis for the tool's code itself to the same criticality level as the system for the code files? You already are developing the PSAC so you include the tool's information there. The tool operational requirement is much lighter in content and in your accomplishment summary you include a summary of how it was implemented.

The tool operational requirement for development tools has two development processes; tool verification and structural coverage analysis of the tool itself. Error condition responses should be included; in other words, inserting errors into the tool to see how it reacts. For the verification tool you only denote functions you are going to use for that tool. That tool may perform many other functions, but you're only intending to use a subset of it. That's the only part to qualify. Show that it works in normal, not abnormal, tests.

## Tool Operational Requirements

**Development Tool**

Tool functionality and operational environment

Operation and installation information

Tool development process

Tool verification & structural coverage analysis

Error condition responses

**Verification Tool**

For components used, tool functionality and operational environment

Operation and Installation Information

Functional test of the tool components used

S283

The overall process for tool qualification in DO-254 is a little different than for DO-178B.

DO-254 says if it's a Level A, B or C design tool or Level A and B verification tool, then you need certain qualification activities. DO-254 gives leniency on service history. DO-178B doesn't. So the table asks, do you need to qualify the tool? If not, you're done. Has the tool been used in these environments before? If yes, then you don't do much more than refer to data items that should already exist. Then do basic tool qualification for a verification tool and design tool qualification if it's a Level A or B design tool. Many design tools for DO-254 are used because much happens in the development of ASICS and PLDs based on the tool sets utilized.

A diagram of the tool qualification decision making process is provided below. Remember to discuss all tool qualification aspects with your DER and do not fail to qualify, or over-qualify, your tools.

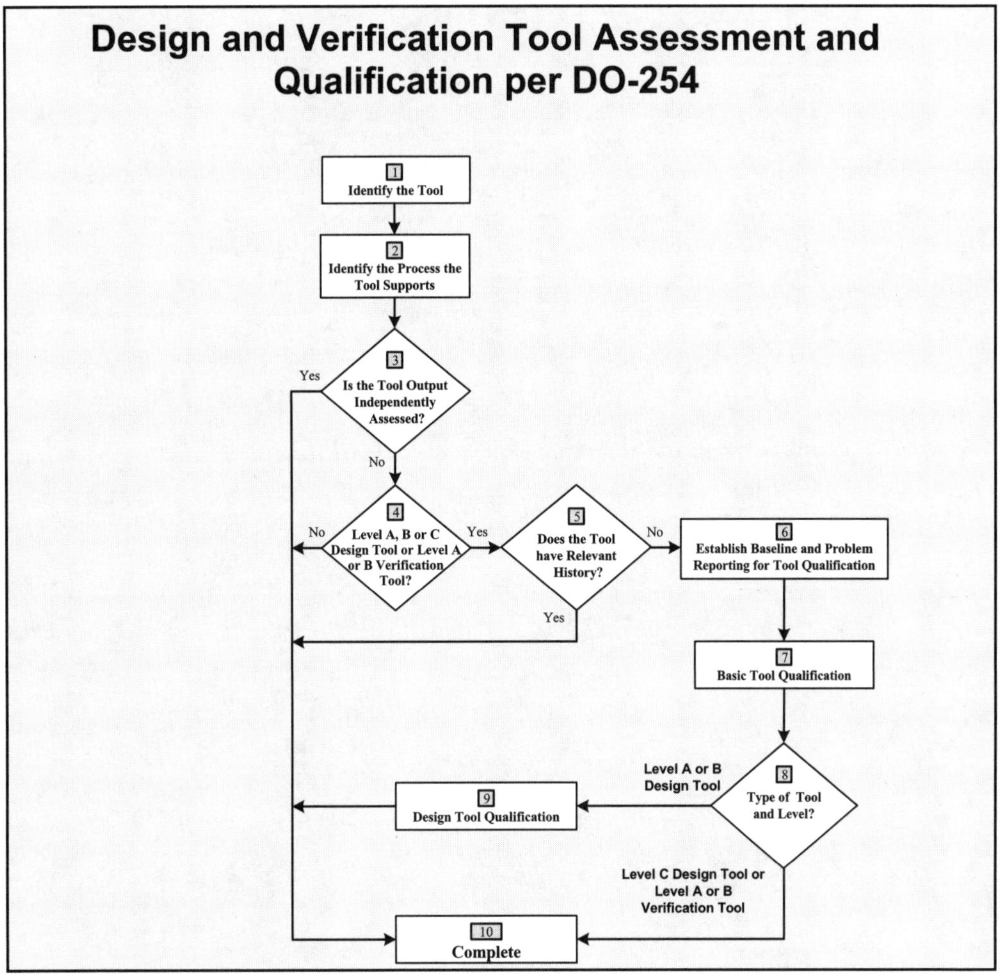
Design and Verification Tool Assessment and Qualification per DO-254
1
Identify the Tool
2
Identify the Process the Tool Supports
3
Is the Tool Output Independently Assessed?
Yes
No
4
Level A, B or C Design Tool or Level A or B Verification Tool?
No
Yes
5
Does the Tool have Relevant History?
No
Yes
6
Establish Baseline and Problem Reporting for Tool Qualification
7
Basic Tool Qualification
8
Type of Tool and Level?
Level A or B Design Tool
9
Design Tool Qualification
Level C Design Tool or Level A or B Verification Tool
10
Complete

# 27. Software Design Aspects

Avionics designs evolved significantly since DO-178B was written in the early 1990's. Hardware and software have become more complex and powerful. At the same time, avionics component and tool vendors greatly expanded their offerings, while competitive pressures and reduced time-to-market posed developmental challenges. DO-178C is being developed as this book is written and new clarification provided for DO-254 certification. What are the top issues needing to be addressed by avionics developers?

**Modern Avionics Design Aspects**

- Multiple criticality levels within a single CPU
- Time partitioning
- Space partitioning
- Resource I/O protection
- ARINC 653 compliance

S286

Key issues evolved in recent years, as shown in the figure above. Multiple criticality systems have been around for a long time. Traditionally, different critical level segments were shared memory, and there were no operating system considerations or assured separation between tasks operating at different criticality levels. With multiple criticality systems, we're trying to execute multiple processes at the same time. If they are on different CPUs, and you can show physical partitioning and completely different software packages, the only concern is communication. There's no problem with shared memory space between the two, and the lower criticality task (which has not been developed or verified as stringently) has no ability to perturb the operation of the higher criticality task. You worry about data being exchanged between them, versus direct corruption of one by the other.

In modern avionics, however, the trend toward weight and power reduction is clear. This means consolidating functionality previously allocated to different LRU's or, at least, different cards within the same LRU. This means a greater propensity to require different tasks with different criticality levels to be executing via the same processor. In this case, there is also the possibility for one task to starve the other of memory or resources. Typically, the culprit is the lower criticality task due its lessened development and verification rigor.

System criticality, then, equals the highest level of criticality within any of the subsystems comprising that system. If one function could lead to a catastrophic effect, the whole software will be Level A because of that one function, even though other functions may only be certified to a lesser level. Why not just develop all subsystems and components to Level A? Cost, pure and simple. Sometimes you want to partition a Level A function from all Level B functions if it's worth doing. But is it worth it? The cost savings between Level B and level A are small so you need to examine your rationale and the potential cost savings. Remember, the partitioning effort is also difficult and you're guilty until proven innocent. If you can show that lower criticality functions won't affect higher software criticality, then you're okay, but you must guarantee it via design, tests and architecture, then prove it to the FAA. To do that, the two most important things are time and space partitioning. No dynamic memory allocation will be allowed.

## Time Partitioning

Time partitioning = deterministic scheduling and execution

Must provide execution overrun detection

No variability in scheduler

Bounded computation time for all system calls

- Prevent usage of system calls which cannot guarantee
- No dynamic data structures
- Memory allocation only at system startup
- Prevent usage of semaphores
  (blocking and synchronization issues)

S288

Good scheduling, deterministic from a time point of view, must provide execution overrun protection. This is particularly important, and challenging if there is a tight time allocation for each system call and its usage. The data structures cannot be dynamically allocated. Memory allocation is only allowed at system start-up. That's why you need a CPU with an MMU (memory management unit) and an RTOS with a memory management system, and an ability to inherently support time and space partitioning. Most of our clients used Green Hills Integrity-178B for this purpose.

### Space Partitioning

Space Partitioning = "Memory Protection"

Utilizes built-in hardware Memory Management Unit (MMU) to enforce execute/read/write permissions

Partition's memory protected from access by another partition

Memory violations are detected and mitigated

System Calls within a partition use that partition's memory allocation to satisfy

No recursion ensures maximum kernel stack size

Resource and I/O protection

S289

**Prevent usage of semaphores due to blocking and synchronization issues**. You must ensure this or you won't be able to prove partitioning. From the space point of view we're looking at memory protection. The CPU provides one level of protection, while the RTOS provides the other. There should be a set allocation of memory so each partition can fully keep within its own allocated memory for all its processing. The only inter-process interaction is via predefined communication protocols, not shared memory. Dynamic allocation is not acceptable. You can't have recursive code or there will be stack overflows. There also must be a method of detecting failures in all systems at all times and for mitigating violators.

**RTOS.** Twenty years ago we didn't often use operating systems. We used a poll loop or, at most, a kernel. In today's critical, complex systems, if you want to develop a more advanced project with upside growth potential, you're Level A or B and have more than five or ten thousand lines of code. Most projects use an RTOS for such implementation. There can be multiple tasks operating asynchronously, reusability, commonality across platforms and a large code base. If there are only several thousand lines of code it's not worthwhile to have a certifiable commercial RTOS unless you have a higher criticality level. If all those are present, you probably want to consider an operating system. The question now is build it yourself, modify one or go COTS?

## RTOS Case Study

**Modern avionics should use a real-time operating system when any of the following is desired**

Multiple tasks operating asynchronously

Reusability

Commonality across platforms

Large (> 5K LOC) source base involved

Higher levels of criticality required

**If *any* of the above are desired, RTOS is suggested. If *all* are desired, an RTOS is mandatory.**

S291

Ten years ago we did everything custom. A typical operating system, though, has over ten thousand lines of code and several million dollars for certification because everything we've discussed has to be done, including structural coverage for MCDC of every line of code. The important features are partitioning for time and space, which means when you exceed a time boundary something happens. The same applies to a space boundary and overwrite. C programmers admit to an occasional bad pointer and space partitioning mitigates that.

Watchdog timers are great but they're limited to system level timing overruns or hung tasks. Modern processors, for example, in the power PC family have built-in partitioning that uses a memory management system in concert with the RTOS to enforce memory constraints. ARINC 653 is a standard for operating systems and portability so, in theory, you have portability. But be careful; in theory you have portability between operating systems but in practice we've rarely seen this accomplished cost effectively.

Operating systems must be certified to the highest level of any software running on them. You don't want to develop your own BSP if you don't have to. Usually, you can buy one from a vendor and he'll modify it for about $30,000 to $50,000, which can be less than spending many personal months or years developing your own.

Questions to ask the operating system vendor: Is the complete certification package available or are they providing more of a do-it-yourself kit? Must you (the buyer) have to create documents yourself and, if so, which ones? Are integrated BSPs available and what's the price, including certification? Is FAA certifiability formally approved already for that version of the RTOS and to the desired criticality level, and how many certifications has that operating system company completed?

These are key questions. Many companies say their operating system is certified, but no operating system is *actually* certified. The most it can ever be is certifiable as a component within the system; remember, it's the system that's certified, not the components. Most RTOS vendors say they've applied for it and certifiability is pending. So be wary of aggressive claims.

**What RTOS Features Necessary for Modern Avionics?**

- Partitioning, time and space
- ARINC 653
- Support for multiple levels of criticality
- Secure I/O protection
- Integration with popular BSP

**Questions to ask commercial COTS RTOS vendor**

- Is the complete certification package, all documents, available?
- Integrated BSPs available?
- FAA certifiability formally approved?
- How many certifications successfully completed?

S293

# 28. Cost Estimation and Metrics

As we near the end of this work, there should be a growing understanding of DO-178B and DO-254, at least well enough to start your project with a proven strategy. Becoming an expert, however, is similar to driving a car; if you haven't worked on an avionics certification project you're at the first stage; with a permit to begin learning. Expertise comes through practice.

We're all concerned with cost and the biggest cost impacts in avionics certification are efficiency and verification. Low efficiency means you will iterate on a solution instead of taking the shortest path...the reason many avionics projects are 50-100% over budget. Verification is a special beast because it consumes about 35% of a DO-178B budget. The reason is structural coverage analysis. How long does it take to verify one line of code? Quite a bit. What are the cost drivers? If you can get the following four items right you may save 30-50%; accurate and detailed requirements, reviews, minimal code changes and efficient testing. The cost goes up rapidly for only slight improvements to the product on life cycle activities.

## DO-178B Cost Metrics/Allocation

CM & QA
10%

DER services
5%

Management
5%

Rqmts development
10%

Design
10%

Code:
25%

Test:
35%

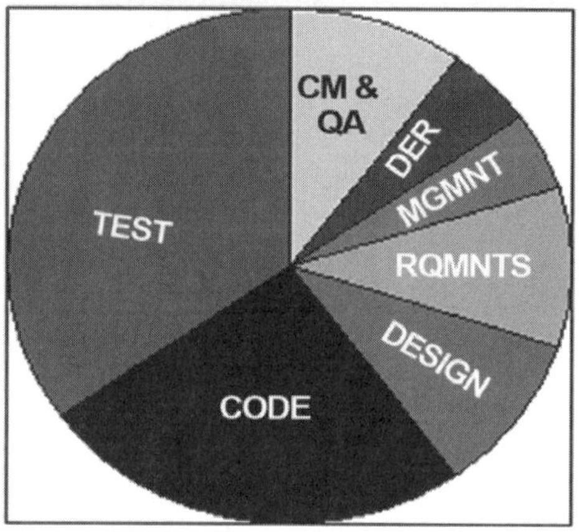


**Primary Cost Drivers**

Accurate & detailed requirements

Accurate & thorough reviews

Minimal code changes

Efficient testing

s296

## Cost Issues

- Cert versus compliance
- Augmenting existing EFW plans for DO-178B (5 key process plans)
- PSAC & SAS
- Application of DO-254
- DO-178B correlation
- DER support
- Formalization of rqmts & traceability
- Automated functional test environment
- Formalization of design methodology
- Structural coverage
- Detailed checklists
- Static code analysis
- Software test tool selection
- Software tool qualification
- RTOS considerations
- BSP certifiability
- Previously existing software
- Gap analysis
- Reverse engineering
- QA upgrades for DO-178B, including audits
- Graphics package/libraries

S299

A popular myth is that DO-178B is expensive. Indeed, even Level D software requires documentation, reviews and basic testing processes. Additionally, configuration management, quality assurance and DER liaison are applied to Level D. Yet, the costs of Level D are hardly more than any non-certified commercial software process. Why? Because Level D is comprised almost entirely of normal industry standard software engineering principles.

Another misconception is that the most significant cost escalation occurs when moving from DO-178B Level B criticality to Level A. Untrue. The cost impact of DO-178B is most significant between Level D and Level C.

The reason is, Level C requires the following, which Level D does not. It results in Level C requiring 30% more budget and schedule:

- Testing low-level software requirements
- Ensuring 100% coverage of all source code statements
- Greater rigor placed on reviews
- In many cases, more rigorous configuration management

Level B requires additional structural coverage (decision-condition, e.g. all branches in the source code), additional independence in reviews and tighter configuration management.

On first glance, then, it seems that Level B should be significantly more expensive, e.g. 50%-70%, than Level C. In theory, it makes sense, but as in many areas of life, common sense overcomes theory. In Level B (and C) there must be detailed low-level software requirements, and they must be thoroughly tested. During requirements-based testing, most (70-90%) of the branches in the source code are already executed and require no additional structural coverage testing if test capture and coverage tools are appropriately used! Therefore, the seemingly significant cost increase associated with Level B versus C structural coverage is already mitigated by requirements-based testing. Also, quality software engineering organizations already incorporate a semi-automated and streamlined process which includes independent reviews and tight configuration management; hence the added cost of those aspects for Level B is largely mollified. The reader is well-advised to undergo upfront DO-178B training and DO-178B Process Improvement to leverage these cost reduction techniques.

Level A is the most critical software level and the most expensive. Still another myth exists, namely *"Level A is very difficult to achieve and will cost at least 30-50% more than Level B."* That's false. Level A imposes yet more structural coverage requirements (MCDC testing), robustness and correlation requirements and slightly more stringent reviews. The most significant cost driver over Level B is the MCDC testing requirement. However, with proper application of modern structural coverage tools, personnel training and thorough requirements-based testing, the added cost for Level A can be largely contained.

It should be apparent that DO-178B does not embody strict requirements, but guidelines. There are three critical processes; planning, development and correctness. The order is planning, correctness, then development. Criticality level is based on the safety assessment. There must be two levels of requirements, high, then low. Traceability embodies top to bottom, then bottom to top. You must show a plan, and adherence to that plan, for transitions between requirements, design, code and test.

Determinism and consistency are primary themes. Audits and proof of conformance are vital. Consider yourself guilty until proven innocent. Independence is not required for Level D. For C---a little bit. For B and A, absolutely. Independence is always recommended, even when you're at Level C and D. What flows are we interested in? Data and control flow. Configuration management must prevent unwarranted changes. QA is independent from everybody, especially development, and enforces the process and adherence to plans via audits.

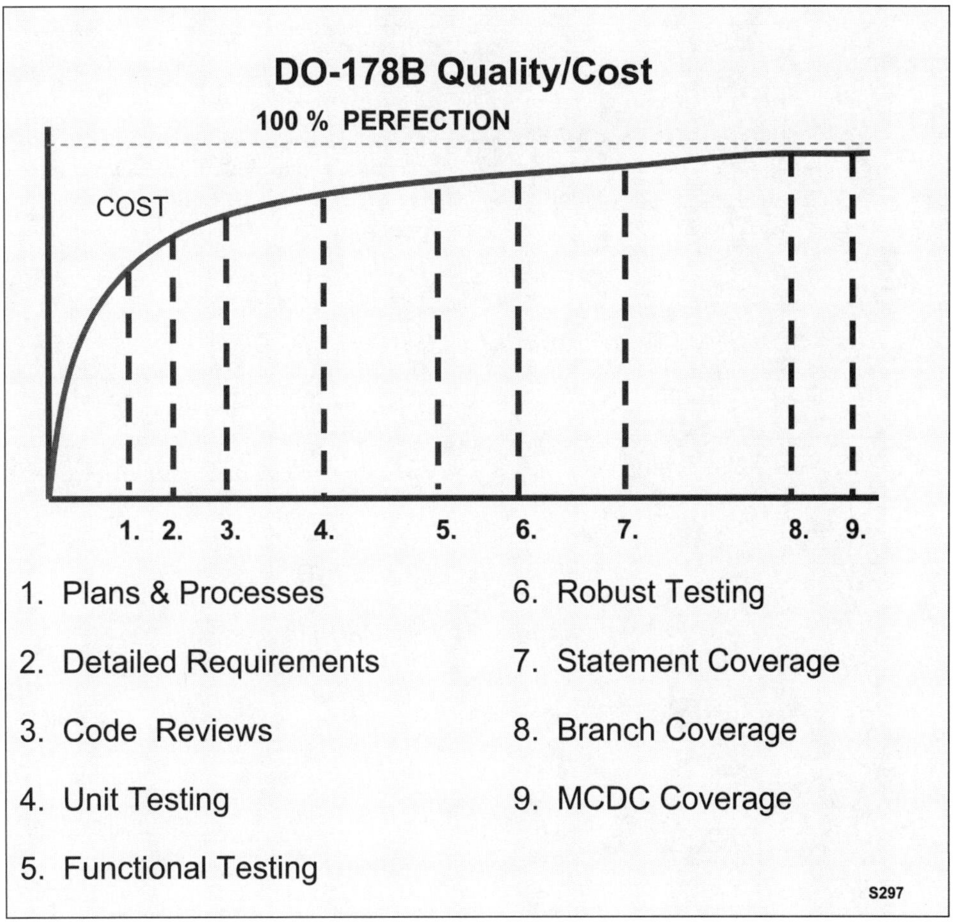


What are the differences between cost effectiveness? Increasing criticality levels entail increasing structural coverage. The added cost of DO-178B should be about 20-40%, depending on the level. Level D should not increase your cost at all; Level C, B and A increase your cost. It should be 60% to 80%, which is the industry average. The primary cost driver is structural coverage. Safe skies everyone!

# Glossary

| | |
|---|---|
| **AC** | Advisory Circular |
| **ACO** | Aircraft Certification Office |
| **AMC** | Alternative Means of Compliance |
| **ARINC** | Aeronautical Radio, Incorporated |
| **ARP** | Aerospace Recommended Practice (SAE) |
| **ASIC** | Application Specific Integrated Circuit |
| **ATP** | Acceptance Test Procedure |
| **BITE** | Built-in Test Equipment |
| **BSP** | Board Support Package |
| **CC** | Control Category |
| **CCA** | Circuit Card/board Assembly |
| **CCB** | Change Control Board |
| **CCDL** | Cross Channel Datalink |
| **CDR** | Critical Design Review |
| **CEH** | Complex Electronic Hardware |
| **CI** | Configuration Index |
| **CM** | Configuration Management |
| **CMM** | Capability Maturity Model |
| **CMMI** | Capability Maturity Model Integration |
| **COTS** | Commercial Off-the-Shelf |
| **CPU** | Central Processing Unit |
| **CSC** | Computer Software Component |
| **CSCI** | Computer Software Configuration Item |
| **CSU** | Computer Software Unit |
| **DEF** | Design Exchange Format |
| **DER** | Designated Engineering Representative |
| **DOD** | Department of Defense |
| **DSP** | Digital Signal Processor |
| **EASA** | European Air Safety Agency |
| **EUROCAE** | European Organization for Civil Aviation Equipment |
| **FAA** | Federal Aviation Administration |
| **FFP** | Functional Failure Path |
| **FHA** | Functional Hazard Assessment |
| **FMEA** | Failure Mode and Effects Analysis |
| **FPGA** | Field Programmable Gate Array |
| **FTA** | Fault Tree Analysis |
| **GPS** | Global Positioning System |
| **HAS** | Hardware Accomplishment Summary |

| | |
|---|---|
| **HDL** | Hardware Design Language |
| **HVP** | Hardware Verification Plan |
| **ICD** | Interface Control Drawing |
| **IFR** | Instrument Flight Rules |
| **JAA** | Joint Airworthiness Authority |
| **LOC** | Lines of Code |
| **LRU** | Line Replaceable Unit |
| **MCDC** | Modified Condition Decision Coverage |
| **OO** | Object Oriented |
| **OOT** | Object Oriented Technology |
| **PDL** | Program Description Language |
| **PDR** | Preliminary Design Review |
| **PHAC** | Plan for Hardware Aspects of Certification |
| **PLD** | Programmable Logic Device |
| **PSAC** | Plan for Software Aspect of Certification |
| **PSSA** | Preliminary System Safety Assessment |
| **QA** | Quality Assurance |
| **QAP** | Quality Assurance Plan |
| **RTCA** | RTCA formerly meant "Radio Technical Commission for Aeronautics," but is now "RTCA, Inc." |
| **RTOS** | Real-Time Operating System |
| **SAS** | Software Accomplishment Summary |
| **SCI** | Software Configuration Index |
| **SCM** | Software Configuration Management |
| **SDP** | Software Development Plan |
| **SECI** | Software Environment Configuration Index |
| **SEI** | Software Engineering Institute |
| **SLOC** | Source Lines Of Code |
| **SOI** | Stage of Involvement |
| **SQA** | Software Quality Assurance |
| **SRS** | Software Requirement Specification |
| **STC** | Supplemental Type Certificate |
| **SWDP** | Software Development Plan |
| **SWVP** | Software Verification Plan |
| **TBD** | To Be Determined |
| **TC** | Type Certificate |
| **TR** | Thrust Reverser |
| **TSO** | Technical Standard Order |
| **TSOA** | Technical Standard Order Authorization |
| **UML** | Unified Modeling Language |
| **UUT** | Unit Under Test |

| | |
|---|---|
| **V&V** | Validation and Verification |
| **VDD** | Version Description Document |
| **VHDL** | Verilog Hardware Design Language |
| **VMC** | Visual Meteorological Conditions |
| **WOW** | Weight on Wheels |